PASS YOUR AMATEUR RADIO EXTRA CLASS TEST – THE EASY WAY
2016-2020 Edition

By: Craig E. "Buck," K4IA

ABOUT THE AUTHOR: "Buck," as he is known on the air, was first licensed in the mid-sixties as a young teenager. Today, he holds an Amateur Extra Class Radio License. Buck is an active instructor and a Volunteer Examiner. The Rappahannock Valley Amateur Radio Club named him Elmer (Trainer) of the Year three times. Buck has successfully led many students through this material.

Email: K4IA@arrl.net

Published by Brigade Press
130 Caroline St. Fredericksburg, Virginia 22401

This and other Easy Way Books by Craig Buck are available at Ham Radio Outlet stores, Amazon, and online retailers:
"Pass Your Amateur Radio Technician Class Test – The Easy Way"
"Pass Your Amateur Radio General Class Test – The Easy Way"
"How to Chase, Work & Confirm DX – The East Way"
"How to Get on HF – The Easy Way"

ISBN 1536916765

Library of Congress Control Number PENDING .5

PASS YOUR AMATEUR RADIO EXTRA CLASS TEST – THE EASY WAY

TABLE OF CONTENTS

TABLE OF CONTENTS

AMATEUR RADIO EXTRA CLASS QUICK SUMMARY

TABLE OF CONTENTS

INTRODUCTION

There are many books to help you study for the amateur radio exams. Most focus on taking you through all the questions and possible answers on the multiple choice test. The problem with that approach is that you must read three wrong answers for every one right answer. That's 2,136 wrong answers and 713 right answers. No wonder people get overwhelmed.

This book is different. There are no wrong answers. I'm going to answer every question on the amateur radio Extra Class exam in a narrative format. **I've put the test questions and answers in bold print to help you focus.** *Hints to help you decipher the questions and answers are in italic print. Cheats to give you a shortcut memory jogger to the answer are also italic.*

The questions are roughly in the order they appear in the question pool. Please excuse any tortured grammar or syntax in the bold print questions and answers. In most cases, I am copying them word-for-word from the exam.

The second part of this book is a Quick Summary. Just the question and the right answer with hints or cheats.

You will pass your Extra Class test, but it won't tell you much about how to operate, chose equipment and put up antennas. Be sure to check out my other books: "***How to Get on HF – The Easy Way***" for detailed instructions to get on the air and "***How to Chase, Work & Confirm DX – The Easy Way.***"

THE TEST

The test is 50 questions out of a possible pool of 712. The pool is large but only about one of fourteen questions in the pool will show up on your exam - one from each test-subject grouping. You won't get 50 questions on one topic. Many of the questions ask for the same information in a slightly different format, so there aren't 712 unique questions.

The Extra test is harder than the General, but you can do it. Don't be intimidated. This book will teach you to recognize the right answer – even if you don't understand the question. Go through this material with the intent to recognize the right answer.

There is no Morse Code required for any class of amateur license. Morse Code is still very much alive and well on the amateur bands and there are many reasons you might want to learn it in the future.[1]

The good news is that we know the exact pool of questions and answers. You want to learn it all, but you only get one question from each group. If there is a question or concept you just can't get, skip it. Chances are it won't be on your test. And, you only need to answer thirty-seven out of the fifty questions to pass. **The minimum passing score is 74%.** (There is your first question and answer).

Most important: when the day comes, take the test even if you feel you are not ready. You will be better prepared than you think and may surprise yourself by passing.

Do you know what they call the person who graduates last in his class from medical school? "Doctor" – the

---

[1] Check out my book "DX – The Easy Way." I discuss the advantages of CW over voice modes in detail.

same as the guy who graduated first in the class. My point is: pass and no one will know the difference.

The questions are multiple choice, so you don't have to know the answer you just have to <u>recognize</u> it. That is a tremendous advantage for the test-taker. However, it's difficult to study the multiple choice format because you can get bogged down and confused by seeing three wrong answers for every correct one. That makes it harder to recognize the right answer when you see it.

<u>The best way to study for a multiple choice exam is to concentrate on the correct answers. That is the focus of this book. When you take the test, the correct answer should jump out at you. The wrong answers will seem strange and unfamiliar as if in a foreign language.</u>

Don't over-think or over-analyze the answers. You should recognize the right answer without thinking. If you don't recognize the answer, try to eliminate the obviously wrong answers and then guess. There is no penalty for guessing wrong, and you greatly increase your odds by eliminating the bogus responses.

Here's the ultimate secret cheat: many times the answer is a matter of common sense and logic, not engineering prowess. You can learn to recognize or figure out the answer without understanding a word of either the question or answer.

The test-day routine is the same as it was for your other tests. You should bring a picture ID, your Social Security Number[2] (not the card, just the number) a

---

[2] Many VEs request you obtain a Federal Registration Number (FRN) from the FCC instead of using your Social Security Number. The FRN is used only for your communications with the FCC. Search "FRN FCC" to get the link to the FCC website where you can register. Then, bring your FRN number to the test session.

THE TEST

pen and pencil and the exam fee (usually around $15). You need to check ahead to get the exact amount and ask if they prefer cash or a check. If you bring a calculator, you must clear the memories. Turn off your cell phone and don't look at it during the exam.

You will fill out a simple application (FCC Form 605). You can get one from the FCC website and fill it out beforehand to save time on test day. The Volunteer Examiner team will select a test booklet. Make sure you ask them for the easy one – that's good for a cheap laugh. They have no idea which questions are in which booklet. You also get a multiple choice, fill-in-the-circle answer sheet. You can take notes and calculate on the back of the answer sheet – not in the booklet. The VEs re-use the booklets.

The VE team will grade your exam while you wait and give you a pass/fail result. Don't ask them to go over the questions or tell you what you missed. They don't know and don't have time to look. When you pass, there is a bit more paperwork, and you walk out with a CSCE – a Certificate of Successful Completion of Examination.

You may not need to take the Extra Class test. Anyone who can demonstrate they once held an Amateur Extra class license which was not revoked by the FCC can get credit. But, you still must pass the current Technician Class test.

If you are already a General, once you have the CSCE, you can operate as an Extra even though you don't yet show up in the FCC database. When you use Extra Class privileges, use the special identifier "AE" after your call sign while waiting for the FCC to post your upgrade on its website. That would be W3ABC/AE. W3ABC stroke/slash/slant AE. *Hint: Awaiting Extra.* After you appear in the database, drop the AE.

HOW YOU SHOULD STUDY

You ace a multiple choice exam by learning to recognize the right answers and eliminating wrong answers. You could study by pouring over the multiple choice questions. That has been the traditional way of most classes and license manuals. The problem with that method is you have to read through three wrong answers to every question. That is both frustrating and confusing. Why study the WRONG answers?

The approach here is that you never see the wrong answers so the right answer should pop out when you see it on the test. You don't need to memorize the whole answer – just enough of it to recognize. Many of the questions and answers are long and involved. They are full of information you don't need to get the correct answer. Sometimes, all you need do is tie two words together. Even if you don't understand any of the question or answer, you can recognize the right answer.

I don't recommend you take practice exams until the week before the test after you feel you have mastered this material. The reason I recommend you wait to take the practice exams is that I don't want you to get confused by seeing all the wrong answers. Take the practice test to build your confidence but study the book. I want you to recognize the right answers first. For free practice exams on the Internet, go to QRZ.com and click on the Resources tab.

Your Amateur Radio license is called a "ticket." Getting a ticket is the beginning of the journey. You can't get on an airplane and start your journey without a ticket. You can't get on the air and start your Amateur Radio journey without a ticket. Get your ticket and the rest of the details will come later.

COMMISSION'S RULES

Your signal is wider than you think so you have to be careful not to get too close to the band edges. That is the lesson of a series of questions.

An SSB phone signal is about 3 kHz wide. **The displayed carrier frequency should be 3 kHz below the upper band edge if using USB.** Your signal will take up 3 kHz above the carrier. **On LSB, the displayed carrier frequency should be 3 kHz above the lower band edge** so your signal will all be above the lower edge.

If you hear a station calling CQ on 14.349 MHz, it is NOT legal to return the call using USB because your sideband will extend beyond the band edge. The 3kHz wide signal would go up to 14.352 MHz which is above the upper band edge of 14.350.
If you hear a station calling CQ on 3.601 MHz LSB, it is NOT legal to return the call because your sideband would extend beyond the band edge.
If you hear a DX station call CQ on 3.500 MHz, it is NOT legal to return the call on the same frequency because one of the sidebands of your CW signal would be out of the band. *Cheat: "Extend beyond the band edge" or "out of the band" is in the correct answers. "Not legal"*

We share 60 meters with other services, so it has some different rules. **The maximum power permitted on 60 meters is 100 watts effective radiated power as compared to a half-wave dipole.**

---

[3] (E1A) refers to the question pool subelement.

The band that uses specific channels rather than a range of frequencies is 60 meters. *Hint: Remember 100 watts, a dipole and channels on 60 meters.*

You set your CW carrier frequency at the center frequency of the channel.

The maximum bandwidth for data emission on 60 meters is 2.8kHz. *Hint: About the same as voice communication.*

Message forwarding stations (like Packet) are usually automated. **If a station forwards a message that violates FCC rules, the control operator of the originating station is accountable for the rules violation.**

If your digital message forwarding station inadvertently forwards a communication that violated FCC rules, you should discontinue forwarding the communication as soon as possible.

You can operate from a ship or airplane, but you must obtain permission from the master of the ship or pilot in command of the aircraft. You need permission from the boss.

On a US registered vessel in international waters, any FCC-issued amateur license will do. No special endorsements required.

Any person holding an FCC-issued amateur radio license or who is authorized for alien reciprocal operation can be in physical control of an amateur station on a vessel or craft registered in the United States. *Hint: Look for "FCC-issued amateur radio license" in both answers."*

STATION RESTRICTIONS (E1B)

A spurious emission is an emission outside its necessary bandwidth that can be reduced or eliminated without affecting the information transmitted. *Hint: Spurious is outside its necessary bandwidth.*

The permitted mean power of any spurious emission is 43 dB below the fundamental emission. *Hint: The question is way too complicated. Just remember spurious emissions must be 43 dB below the fundamental emission.*

An amateur station or antenna might be restricted if the location is of environmental, importance, or significant in American history, architecture or culture.
Before placing an amateur station within an officially designated wilderness area or wildlife preserve or an area listed in the National Register of Historical Places, you must submit an Environmental Assessment to the FCC.
Cheat: Look for the word "environmental" in the answer. Everyone cares about the environment.

Before installing an antenna at a site near a public use airport, you may have to notify the Federal Aviation Administration and register it with the FCC.

If you are within 1 mile of an FCC monitoring facility, you must protect it from harmful interference.

The National Radio Quiet Zone is an area surrounding the National Radio Astronomy Observatory. *Cheat: "National is in both the question and answer.*

The highest modulation index permitted at the highest modulation frequency for angle modulation below 29 MHz is 1. Angle modulation is FM. Below 29 MHz, it has to be narrow-band FM. The answer is 1.

If an amateur station's signal causes interference to <u>domestic broadcast</u> reception, the FCC may place limitations to avoid transmitting during certain hours on frequencies that cause the interference. The FCC can impose "quiet times."

RACES is the Radio Amateur Civil Emergency Service and is part of a protocol established by FEMA and the FCC. **Any FCC-licensed amateur station certified by the responsible civil defense organization for the area served may be operated under RACES rules.** *Hint: RA<u>C</u>ES <u>c</u>ertifies.*

The frequencies authorized for amateur stations under RACES rules are all amateur service frequencies authorized to the control operator. Nothing special.

CONTROL (E1C)

A remotely controlled station is one controlled indirectly through a control link. You are twiddling the knobs from afar.

A control operator must be present at the control point.

The maximum allowed duration of a remotely controlled station's transmissions if its control link malfunctions is 3 minutes. It must time-out after 3 minutes.

Local control is direct manipulation of the transmitter by a control operator. You are sitting in front of the equipment twiddling the knobs.

Automatic control uses devices and procedures so the control operator does not have to be present. A repeater is on automatic control.
The control operator responsibilities under automatic control are different from under local control because the control operator is not required to be present at the control point.

An automatically controlled station may never originate third party communications.

Automatic control is allowed below 30 MHz on 29.500 – 29.700 MHz. Only in the top of the 10-meter band.

Auxiliary, repeater or space stations are the types allowed to automatically retransmit signals of other amateur stations. *Hint: NOT beacons. Beacons don't retransmit.*

The control operator of an auxiliary station can be a Technician, General, Advanced or Extra class operator. *Hint: Look for the answer with "Technician."*

IARP is an International Amateur Radio Permit that allows amateurs to operate in certain countries of the Americas. *Hint: Remember it stands for International Amateur Radio Permit.*

CEPT is an agreement for US amateurs to operate in European countries.
To operate in accordance with the CEPT rules you must bring a copy of FCC Public Notice DA 11-221. It lists the countries and rules in several languages. *Cheat: Remember "FCC Public Notice."*

A Canadian license holder is allowed the same privileges in the US, not to exceed the US Extra Class privileges.

Communications incidental to the purpose of the amateur service and remarks of a personal nature are the types of communication that may be transmitted to amateur stations in foreign countries. Same as domestic.

SATELLITES (E1D)

The amateur satellite service is a radio communications service using amateur radio stations on satellites.

Telemetry is one-way transmission of measurements. *Hint: Telemetry is metering.*

A telecommand station is a station that transmits communications to initiate, modify or terminate functions of a space station. *Hint: It commands the space station to do something.*

A space station must be capable of terminating transmissions by telecommand when directed by the FCC. It must have an "off" switch.

Any amateur station designated by the space station licensee is eligible to be a telecommand station subject to the privileges of the control operator license class. *Hint: You need permission from the boss, the space station licensee.*

An Earth station is an amateur station within 50 km of the Earth's surface intended for communications with amateur stations by means of objects in space. *Cheat: Ditch the complicated answer. The correct answer is the only one that mentions 50 km.*

COMMISSION'S RULES

Any class of amateur license with appropriate operator privileges is authorized to be the control operator of a space station. Many of the astronauts are Technicians.

The HF bands authorized for space stations are 40 m, 20 m, 17 m, 15 m, 12 m, and 10 m bands. *Hint: All of 40-10 except 30 m.*

On VHF, only 2 meters is available for space stations.

On UHF, 70 cm and 13 cm are available for space stations.

EXAMINERS (E1E)

Count yourself fortunate that you do not have to travel to an FCC Field Office to take your test. The FCC turned testing responsibilities over to Volunteer Examiner Coordinators in 1984. There are about 14 Volunteer Examiner Coordinator organizations, and they jointly write the question pools and administer the system. The folks you see on test day are Volunteer Examiners.

The questions for all written U.S. amateur license examinations are listed in a question pool maintained by all the VECs.

The Volunteer Examiner Coordinator is an organization that has entered into an agreement with the FCC to coordinate amateur radio license examinations. *Hint: A coordinator coordinates.*

The Volunteer Examiner accreditation process is the procedure by which a VEC confirms that the VE applicant meets FCC requirements. *Hint: A VE is accredited by a VEC, not the FCC.*

It takes at least three qualified VEs to administer an Element 4 amateur operator license examination.
When an examinee scores a passing grade, three VEs must certify that the examinee is qualified for the license grant and that they have complied with the VE requirements. *Cheat: Ditch the overcomplicated question and answer. Look for "three VEs" and that is the answer to both questions.*

If the examinee does not pass the exam, the VE team will return the application form to the examinee. You get the application back, not the fee.

After administering a successful examination, the VEs must submit the application to the coordinating VEC according the VEC's instructions. The VEC loads the data in the FCC database.

A VEC can conduct an exam session remotely using a real-time video link and the Internet to connect the exam session to the observing VEs. *Hint: Tie "remote session" and "Internet."*

If a candidate fails to comply with the examiner's instructions, the examiner should warn him that continued failure to comply will result in termination of the examination.

A VE may not administer an examination to someone who is a relative of the VE as listed in the FCC rules. *Hint: Friends and employees are OK.*

A VE who fraudulently administers or certifies an examination can lose his amateur station license and amateur operator license. No fine or jail time – loss of license is worse!

COMMISSION'S RULES

Each administering VE is responsible for the proper conduct and necessary supervision during an amateur radio license examination session.

VEs and VECs can be reimbursed for out-of-pocket expenses such as preparing, processing, administrating and coordinating an examination.

MISCELLANEOUS RULES (E1F)

Spread spectrum transmissions are those that change frequency deliberately to decrease interference and assure privacy. The result is a wide signal so it is limited to the higher and wider bands. **Spread spectrum transmissions are permitted on 222 MHz and above.**

If you are going to use spread spectrum, the following conditions apply:
- **Must not cause harmful interference**
- **Must be in an area regulated by the FCC or in a country that permits SS emissions**
- **Must not be used to obscure the meaning**
- **All of the choices are correct**
  Hint: Lots of conditions so all are correct

The maximum transmitter peak envelope power for transmitting spread spectrum is 10 watts.

Amplifiers must be FCC certified to protect against use by CBers and to assure spectrum purity. **A dealer may sell an external RF power amplifier capable of operation below 144 MHz if it has not been granted FCC certification if it was purchased in used condition from an amateur operator and is sold to another amateur operator for use at that operator's station.** *Hint: Too complicated! Just remember amateur to amateur is OK.*

An amplifier must satisfy the FCC's spurious emissions standards when operated at 1500 watts or its full output power to qualify for a grant of FCC certification.
Hint: To get certification, satisfy the FCC.

**Line A is a line roughly parallel to and south of the US-Canadian border.
Amateur stations in the US may not transmit on 420 MHz – 430 MHz if they are above the line.**

A Special Temporary Authority may be issued by the FCC to provide for experimental amateur communications. These usually allow for operation outside the normal amateur frequencies.

Communications transmitted for hire or material compensation, except as otherwise provided in the rules are prohibited. *Hint: "For hire" is prohibited. You can't be paid for being a Ham.*

Amateur radio isn't for business. Here's an example: I'm on my way over to my friend Bob's house to help him with an antenna. I call him on the radio and ask if he would like me to bring over pizza. He says, sure – whatever kind you like. Another ham, who also happens to own a pizza parlor, hears me and calls. "Hey Buck, what kind of pizza do you want. I'll have it ready for you." That's a no-no. Same pizza but a different result.

Extra Class – The Easy Way   Page 21

OPERATING PROCEDURES (E2)

AMATEUR RADIO IN SPACE (E2A)

The ascending pass for an amateur satellite is from south to north. *Hint: "Ascending." Think of it as going up from the South Pole to the North.*

A descending path is from north to south. *Hint: "Descending" from the North Pole to South. The opposite of above.*

The orbital period of an Earth satellite is the time it takes for it to complete one revolution around the Earth. *Hint: Orbit = revolution.*

A satellite's mode is the uplink and downlink frequency bands.
The letters in a satellite's mode designator specify the uplink and downlink frequencies.
If a satellite is operating in U/V mode, it means it is in UHF on the uplink and VHF on the downlink. **U/V mode means the satellite would receive on 435 MHz – 438 MHz.** *Hint: That is UHF.*

A linear transponder can relay
FM and CW
SSB and SSTV
PSK and Packet
All of these choices are correct
Hint: The transponder is linear, so it can relay any type of signal.

The effective radiated power to a satellite which uses a linear transponder should be limited to avoid reducing the downlink power to all other users. A satellite receiving a very strong signal assumes a very good connection and will reduce its own power to conserve energy.

little 23 *small 13* **OPERATING PROCEDURES**

The terms L band and S band refer to the 23-centimeter and 13-centimeter bands.

A satellite signal may exhibit rapid repeated fading because the satellite is spinning. The antenna points away from Earth.

To minimize the effect of spin modulation and Faraday rotation, use a circularly polarized antenna. *Cheat: Spin in circles.*

To predict the location of a satellite at a given time, use calculations using Keplerian elements for the specified satellite. *Hint: Kepler developed formulas for describing an orbiting body.* ✓

A satellite that stays in one position in the sky is geostationary. *Hint: If it stays in one place it is stationary.*

TELEVISION PRACTICES (E2B)

Fast-scan (NTSC) television transmits 30 frames per second.

Fast-scan (NTSC) television transmits 525 lines per frame.

NTSC is the video standard used by North American Fast Scan ATV stations[4].

An interlaced scanning pattern is generated in a fast-scan (NTSC) television by scanning odd numbered lines in one field and even numbered lines in the next. *Hint: Odd and even are interlaced.* ✓

---

[4] ATV is amateur television.

Blanking of a video signal is turning off the scanning beam while it is traveling from right to left or from bottom to top. You wouldn't want the beam to leave a trace as it returns.

Vestigial sideband modulation is amplitude modulation in which one complete sideband and a portion of the other are transmitted. *Hint: Vestigial means a small remnant – a portion of the other sideband.*

The advantage of vestigial sideband for standard fast-scan television is that it reduces bandwidth while allowing for simple video detector circuitry. Because only part of the other sideband is transmitted, it reduces bandwidth. *Hint: "Reduces bandwidth." (a good thing)*

The signal component that carries color information is called chroma. *Hint: Chromatic means of, relating to, or produced by color.*

A common method of transmitting accompanying audio with amateur fast-scan television is:
- **Frequency-modulated sub-carrier**
- **A separate VHF or UHF audio link**
- **Frequency modulation of the video carrier**
- ✶ **All of these choices are correct**
Hint: All of these are ways to pass audio.

To decode SSTV, a receiver with SSB capability and a suitable computer is all the hardware needed. SSTV is slow-scan TV.

DRM is a digital codec used to send radio signals in less bandwidth and at higher quality. The computer decodes the signal. **An acceptable bandwidth for DRM-based voice or SSTV digital transmissions is 3 KhZ.** That is about the same as SSB and half of a standard AM signal.

The Vertical Interval Signaling (VIS) code is sent as part of an SSTV transmission to identify the SSTV mode being used. *Hint: It is signaling the mode.*

Analog SSTV images are typically transmitted by varying tone frequencies representing the video transmitted using on single sideband. Listen on 14.230 MHz to hear SSTV.
Brightness of the picture is encoded by the tone frequency.
Receiving equipment is signaled to begin a new line by specific tone frequencies.
Hint: "Tone frequencies" are in all the answers.

Amateur slow-scan TV uses 128 or 256 lines per frame. That is half of fast-scan (525 lines).

Special frequency restrictions for slow-scan TV are that they are restricted to phone band segments and their bandwidth can be no greater than a voice of the same modulation type. *Hint: They are restricted to voice band segments.*
The approximate bandwidth of slow-scan TV is 3 kHz. The same as DRM, the same as SSB.

FM ATV transmissions can be heard on 1255 MHz.

OPERATING METHODS: CONTEST AND DX (E2C)

During contesting, operators are permitted to make contacts even if they do not submit a log. You can contest all you want and never bother to submit a log to the contest sponsor. But, you should send in a log to help with the score checking, and you might win an award without realizing it.

OPERATING PROCEDURES

The standard for submission of electronic logs is Cabrillo format. Your computer logging program will convert your file to Cabrillo format for uploading to the contest sponsor.

Self-spotting is the generally prohibited practice of posting your own callsign and frequency on the spotting network. You can't spot yourself.

Amateur contesting is not allowed on 30 meters. In fact, it is not allowed on any of the WARC bands.

During a VHF/UHF contest, you would expect to find the highest level of activity in the weak signal segment of the band with most of the activity near the calling frequency. *Hint: Most people will listen for weak signals and listen near the calling frequency.*

A mesh network is interconnected wireless points. **The type of transmission used for ham radio mesh networks is spread spectrum in the 2.4 GHz band.** Hams are repurposing inexpensive wireless routers to create mesh networks.

The equipment commonly used to implement a ham radio mesh network is a standard wireless router running custom software.

A DX QSL manager handles the receiving and sending of confirmation cards for a DX station.[5]

The U.S. QSL bureau system is used for contacts between a U.S. station and a non-U.S. station. The U.S. bureau system does not handle domestic contacts.

---

[5] Lots more details in my book "How to Chase, Work & Confirm DX – The Easy Way"

Many DX stations operate "split," listening on a different frequency from their transmissions.

A DX station might state they are listening on a different frequency:

Because they are transmitting on a frequency prohibited to some of the responding stations.

To separate the calling stations from the DX station.

To improve the operating efficiency by reducing interference.

All of these choices are correct.

When attempting to contact a DX station during a contest or pileup, you would generally identify by sending your full call once or twice. No partial calls, no grid squares, no repetitive identifying.

If DX becomes too weak to copy a few hours after sunset, switch to a lower frequency HF band. Follow the propagation.

When a U.S. licensed operator is operating a remote control transmitter in the U.S., no additional indicator is required (when identifying).

OPERATING METHODS: VHF/UHF AND DX (E2D)

The digital mode especially designed for meteor scatter is FSK441.

A good technique for making meteor scatter contacts is:

15 second timed transmission sequences with stations alternating based on location.

Use of high-speed CW or digital modes.

Short transmission with rapidly repeated calls signs and signal reports.

All the choices are correct.

Hint: Lots of good techniques, so all are correct.

OPERATING PROCEDURES

The digital mode especially useful for EME communications is JT65. EME is Earth-Moon-Earth, and since neither moves very fast, you can use a slow mode, like JT65, designed for extremely weak signals.

JT65 improves EME communication because it can decode signals many dB below the noise using FEC. *Hint: Too much information. Remember JT65 decodes signals below the noise.*
FEC is forward error correction.

The advantage of using JT65 coding is the ability to decode signals which have a very low signal to noise ratio.

The type of modulation used for JT65 contacts is multi-tone AFSK. AFSK is Audio Frequency Shift Keying. *Hint: Remember JT65 uses audio tones.*

The method of establishing EME contacts is time synchronous transmissions alternatively from each station. *Hint: To establish contact you need to know when to listen (time synchronous).*

JT65 contacts are organized by alternating transmissions at 1-minute intervals. *Hint: Alternating transmission*

Digital store and forward functions on an Amateur Radio satellite store digital messages in the satellites for later download by other stations. *Hint: The name says it all.*
The technique used by low Earth orbiting digital satellites to relay messages around the world is called store-and-forward. *Hint: Too much information! You relay messages by storing and forwarding them.*

APRS is Automatic Packet Reporting System. A GPS interfaces with your radio to send and receive position reports.

The data used by an APRS network to communicate your position is latitude and longitude. *Hint: Latitude and longitude show your position.*

The digital protocol used by APRS is AX.25.

The packet frame used to transmit APRS beacon data is unnumbered information. *Hint: The only answer with "information" Tie "data" and "information."*

An APRS station can help support public service communications because a GPS unit can automatically transmit information to show a mobile station's position during the event.

OPERATING METHODS: HF DIGITAL (E2E)

A common type of data emissions below 30 MHz is FSK. Frequency Shift Keying. This is used for RTTY (teletype). FSK is not the only type but it is the only answer that is correct. *Hint: Recognize that FSK is a digital mode.*

If one of the ellipses in an FSK crossed-ellipse display suddenly disappears, selective fading has occurred. *Hint: The ellipse faded away.*

The difference between direct FSK and audio FSK is direct FSK applies the data signal to the transmitter VFO. Diddling the frequency changes the tone on the receiving end. Audio FSK feeds the audio tone into the microphone. From there, it is transmitted as SSB.

OPERATING PROCEDURES

If you are using Automatic Link Enable (ALE), you are using automatic control. *Hint: "Automatic" is in the question and answer.*

The digital mode that does NOT support keyboard-to-keyboard operation is Winlink. Winlink is an Internet/radio system for sending Email. It is not a chat room.

The most common data rate for HF packet is 300 baud.

A properly modulated MFSK16 signal is 316 Hz wide. MFSK is multiple frequency shift keying.

The HF digital mode used to transfer binary files is PACTOR. PACTOR sends and receives digital information using radio and the Internet.

PSK31 has the narrowest bandwidth (31Hz) **PSK31 uses variable-length coding for bandwidth efficiency.** The letters have different lengths. Capital letters take twice as long to send so don't use all caps.

If you are trying to initiate contact with a digital station on a clear frequency but are unsuccessful, the reason might be:
Your transmit frequency is incorrect.
The protocol version you are using is not supported by the digital station.
Another station you are unable to hear is using the frequency.
All of these choices are correct.
Hint: Lots of reasons it might not work. If you recognize two, "all of the above" is the answer.

RADIO WAVE PROPAGATION (E3)

ELECTROMAGNETIC WAVES (E3A)

The maximum separation between two stations operating Moon bounce is 12,000 miles if the Moon is visible by both stations. *Hint: Forget the miles. The only way Moon bounce works is if the Moon is visible by both stations. Common sense.*

Libration fading of an EME signal is a fluttery irregular fading. Libration is a perceived oscillating motion of two orbiting bodies. It would be irregular.

When scheduling an EME contact you would want the Moon to be at perigee. Perigee is nearest to the Earth. Your signal doesn't have to travel as far.

Hepburn maps predict the probability of tropospheric propagation.

Tropospheric propagation often occurs along weather related structures of warm and cold fronts.

Rain can support microwave propagation as radio waves bounce off the water droplets. **The rain must be within radio range of both stations for microwave propagation via rain scatter.** *Hint: If the radio wave can't reach the rain, the rain can't reflect it.*

Atmospheric ducts capable of propagating microwave signal often form over bodies of water. *Hint: Microwaves bounce off water vapor.*

Meteor scatter is formed by free electrons in the E layer. *Cheat: Meteor, free, electrons = lots of Es.*

PROPAGATION

Meteor scatter is most suited to 28 MHz – 148 MHz. *Hint: Six meters is popular for meteor scatter and this answer is the only one that includes 50MHz (six meters)*

The atmospheric structure that can create a path for microwave propagation is temperature inversion. The signal bounces between layers of different temperatures.

The typical range for tropospheric propagation is 100 – 300 miles.

Auroral activity is caused by interaction in the E layer of charged particles from the Sun with the Earth's magnetic field. *Hint: Aurora is associated with the Earth's magnetic field.*

The best mode for aurora propagation is CW. *Hint: CW has a 10 dB advantage over SSB and is almost always the best mode for all propagation.*

From the contiguous 48 states, you should point your antenna north to take maximum advantage of aurora propagation. *Hint: Aurora is associated with the Earth's magnetic field. The activity is strongest over the North magnetic pole.*

An electromagnetic wave consists of an electric field and a magnetic field. *Hint: The obvious answer.*

Electromagnetic waves travelling in free space are described as changing electric and magnetic fields propagate the energy. *Hint: The waves propagate.*

Circularly polarized electromagnetic waves are waves with a rotating electric field. *Hint: Circles rotate.*

TRANSEQUATORIAL PROPAGATION (E3B)

Transequatorial propagation is between two mid-latitude points at approximately the same distance north and south of the magnetic equator. *Hint: If is it is <u>trans</u>equatorial, it must cross the equator.*

The approximate maximum range for transequatorial propagation is 5,000 miles.

The best time for transequatorial propagation is afternoon or early evening. *Hint: Give the sun all day to charge up those electrons.*

The terms extraordinary and ordinary waves refers to independent waves created in the ionosphere that are elliptically polarized. Linearly polarized waves that split into ordinary and extraordinary become elliptically polarized. *Hint: E&O waves are elliptically polarized.*

Long path propagation is supported by 160 – 10 meters. *Hint: All HF bands support long path.* **The band that most frequently provides long-path propagation is 20 meters.** *Hint: The band usually open for DX is 20 meters.*

An echo on the received signal of a distant station could be accounted for by receipt of the signal by more than one path. *Hint: The signals on additional paths would arrive at slightly different times and sound like an echo.*

HF propagation along the terminator between daylight and darkness is called "gray-line." *Hint: The line between day and night is gray.*

The cause of gray-line propagation is at twilight and sunrise, D-layer absorption is low while E

and F layer propagation remains high. *Hint: Gray-line occurs at twilight and sunrise.*

Sporadic E propagation is most likely to occur around the summer solstice. *Hint: Sporadic E usually happens in the summer.*
Sporadic E can occur any time during the day. *Hint: It is a daytime phenomenon.*

The primary characteristic of chordal hop propagation is successive ionospheric reflections without an intermediate reflection from the ground. Chordal hop means the signal bounces across the inside of the sphere formed by the ionosphere and doesn't hit the Earth until it finally comes back down at the receiver.

Chordal hop is desirable because the signal experiences less loss compared to normal skip propagation. *Hint: Less loss is a good thing.*

RADIO-PATH HORIZON (E3C)

Ray tracing refers to modeling a radio wave's path through the ionosphere. *Hint: You trace the wave's path.*

A rising A or K index indicates increasing disruption of the geomagnetic field. *Hint: The index rises with geomagnetic disturbances.*

When the A or K index is elevated, polar paths are most likely to experience high levels of absorption. *Hin: It is a geomagnetic disturbance and that would be strongest around the poles.*

The value of Bz represents the direction and strength of the interplanetary magnetic field. *Hint: B and z represent two things, direction and strength. The other answers are only one thing.*

The orientation of Bz that increases the likelihood incoming particles from the Sun will cause disturbed conditions is southward. *Hint: If the waves are headed southward (to the south), they will hit the North magnetic pole and cause disruptions.*

You learned on the Technician Class test that UHF/VHF radio waves can bend slightly over the horizon. **The radio-path horizon distance exceeds the geometric horizon because of downward bending due to density variations in the atmosphere.** *Hint: The air bends the wave over the horizon.*

The radio horizon can exceed the geometric horizon by about 15% of the distance. *Cheat: Not a lot and this is the lowest answer.*

Solar flares are categorized by a letter. **The greatest solar flare intensity is Class X.** *Hint: X and in "extreme."*

An X3 flare is twice as intense as an X2 flare.

A sudden rise in radio background noise might indicate a solar flare has occurred.

Space weather also gets a letter. **The term G5 means an extreme geomagnetic storm.** *Hint: "G" as in "geomagnetic."*

The 304A solar parameter measures UV emissions at 304 angstroms, correlated to the solar flux index. *Cheat: 304A as in "Angstroms."*

VOACAP software models HF propagation.
The Voice of America Coverage Analysis Program was developed to help VOA schedule programming to different parts of the world.

PROPAGATION

As the frequency is increased the maximum distance of ground wave propagation decreases. The ground attenuates higher frequencies more.

The type of polarization best for ground-wave propagation is vertical. *Hint:* *AM radio towers are vertical.*

AMATEUR PRACTICES (E4)

TEST EQUIPMENT (E4A)

The parameter that determines the bandwidth of an oscilloscope is sampling rate. *Hint: The bandwidth is determined by the rate the oscilloscope can sample the signal. Higher rate = higher bandwidth.*

A spectrum analyzer would display RF amplitude and frequency. *Hint: It is analyzing the spectrum and shows frequency and strength of the signal.*

To display spurious signals and/or intermodulation distortion products in an SSB transmitter, use a spectrum analyzer. *Hint: Too complicated! To display spurious signals, use a spectrum analyzer to see them.*

To measure intermodulation distortion in an SSB transmitter, modulate with two non-harmonically related audio frequencies and observe the output on a spectrum analyzer. *Hint: You modulate with audio and you measure distortion with a spectrum analyzer.*

An important precaution to follow when connecting a spectrum analyzer to a transmitter output is to attenuate the transmitter output going to the analyzer. *Hint: Think about it, you wouldn't want to pour a lot of power into a sensitive piece of test equipment.*

The upper-frequency limit for a computer soundcard-based oscilloscope is determined by the analog-digital conversion speed of the

soundcard. *Hint: The speed of the soundcard determines the limits.*

That speed is called the sample rate. **The highest frequency signal that can be digitized without aliasing is one-half the sample rate.**

The advantage of a digital vs. analog oscilloscope is:
Automatic amplitude and frequency numerical readout.
Storage of traces for future use.
Manipulation of time base after trace capture.
All of these choices are correct.
Hint: Digital has many advantages over analog.

The effect of aliasing in an oscilloscope is false signals are displayed. *Hint: An alias sends a false signal.*

The advantage of an antenna analyzer over an SWR bridge is antenna analyzers do not need an external RF source. *Hint: An antenna analyzer generates its own RF.*

The instrument best for measuring the SWR of a beam antenna is an antenna analyzer.

An antenna analyzer should be connected directly to the feed line. Antenna analyzers have a coax jack to plug in the coax.

To display multiple digital signal states simultaneously, use a logic analyzer. Signal states are high/low or on/off and are used in logic circuits.

When using an oscilloscope probe it is good practice to keep the signal ground connection as

short as possible. *Hint: A short ground lead has less effect on the circuit being measured.*

Calibrating an oscilloscope probe is called "compensating." **Compensation of an oscilloscope probe is typically done by displaying a square wave and adjusting the probe, so the horizontal lines are as flat as possible.** *Hint: A square wave shows nice flat lines.*

A prescaler on a frequency counter divides a higher-frequency signal so a low-frequency counter can display it. *Hint: It scales the signal before it is measured.*

The advantage of a period-measuring frequency counter over a direct count is that it provides improved resolution of low-frequency signals within a comparable time period. *Hint: Think of period-measuring as averaging. It will be more accurate than an instantaneous reading.*

MEASUREMENTS (E4B)

The accuracy of a frequency counter is most affected by the time base accuracy. *Hint: The base must be accurate.*

If a frequency counter has an accuracy of +/- 1.0 ppm and reads 146.520 MHz, the most the actual frequency could differ from the reading is 146.52 Hz. *Hint: "PPM" is "parts per million." The answer is one part per million or one-millionth of the reading.*
Same question but the accuracy is +/- .1 ppm. The answer is one-tenth of one million or 14.652 Hz.
Same question but the accuracy is +/- 10 ppm. The answer is ten times or 1465.20 Hz.

The advantage of using a bridge circuit to measure impedance is it is very precise in obtaining a signal null.

The power absorbed by the load when the forward power is 100 watts and the reflected power is 25 watts is 75 watts. *Hint: Simple math. 100 watts went out and 25 came back so, 75 watts must have stayed in the load.*

The subscripts of S parameters represent the port or ports at which the measurements are made. *Cheat: S as in "ports."*[6]

The S parameter equivalent to forward gain is S21. *Cheat: We looked forward to turning 21.*

The S parameter that represents return loss or SWR is S11. *Cheat: A 1:1 SWR is good.*

A characteristic of a good DC voltmeter is a high impedance input. That means it won't load down the circuit and cause false readings.

The significance of voltmeter sensitivity expressed in ohms per volts is a full-scale reading of the voltmeter multiplied by its ohms per volt will indicate the input impedance. *Hint: Algebra. Volts x ohms/volt = ohms. Ohms indicate impedance.*

The current reading on an RF ammeter in series with the antenna feed line will increase if there is more power going into the antenna. *Hint: More amps = more power.*

---

[6] Sometimes there are random questions that mean nothing to you. The only way to recognize the correct answer is with a cheat. These cheats have nothing to do with a correct answer but will help you recognize one. They are my way of getting back at the test designers for asking these types of questions.

To measure intermodulation distortion (IMD) in an SSB transmitter, modulate with two non-harmonically related audio frequencies and observe the output on a spectrum analyzer. *Hint: You modulate with audio, and you measure distortion with a spectrum analyzer.*

If a dip meter is too tightly coupled to the tuned circuit being checked, a less accurate reading results. *Hint: A grid dip meter is used to check the resonate frequency of a circuit. It is a meter and if something is wrong with your test procedure (too tightly coupled), you will get bad readings.*

A relative measurement of the Q for a series-tuned circuit is the bandwidth of the circuit's frequency response. Higher "Q" means narrower bandwidth. *Hint: You measure Q by measuring the frequency response. Forget about "series tuned."*

To calibrate a RF vector network analyzer you use a short circuit, open circuit and 50-ohm loads. *Hint: You calibrate at the two extremes and the "normal" 50-ohm load of coax.*

RECEIVER PERFORMANCE CHARACTERISTICS (E4C)

The effect of excessive phase noise in the local oscillator of a receiver is it can cause strong signals on nearby frequencies to interfere with the reception of weak signals. *Hint: Excessive noise causes interference.*

The portions of a receiver that can be effective in eliminating image signal interference are a front-end filter or pre-selector. *Cheat: The question asks about portions (plural) and only one answer has more than one. Read the question carefully. Sometimes the question gives away the answer.*

The term for the blocking of one FM phone signal by another stronger signal is capture effect.
Hint: FM receivers capture the strongest signal.

The noise figure of a receiver is defined as the ratio in dB of the noise generated by the receiver to the theoretical minimum noise. Hint: the question asks about the noise figure of a receiver. The answer must mention noise generated by the receiver. Again, the question gives away the answer.

The value of -174 dBm/Hz noise floor represents the theoretical noise at the input of a perfect receiver at room temperature. *Cheat:* Guaranteed to be the only answer that has "room temperature."

The MDS of a receiver is the minimum discernible signal. It is a measure of receiver sensitivity.

The primary source of noise heard from a receiver with an antenna connected is atmospheric noise.

An SDR is a software defined radio. Analog signals are converted to digital ones and zeros. Then, computer power sorts them out. The analog-to-digital converter is the heart of an SDR.
An SDR receiver is overloaded when signals exceed the maximum count value of the analog-to-digital converter.

The largest effect on an SDR receiver's linearity is the analog-to-digital converter sample width. Hint: Look for analog-to-digital converter in the answer. An SDR needs one.
Missing codes in the analog-to-digital converter will cause distortion. Hint: Missing data would cause distortion.

Conventional super-heterodyne receivers convert the incoming signal to a fixed intermediate frequency (IF) and feed it through filters designed for that frequency. **A good reason for selecting a high frequency for the IF is it is easier for the front end circuitry to eliminate image responses.** *Hint: Mixing "up" to a higher frequency puts the images farther away from the intermediate frequency.*

The mixing can produce images. If the intermediate frequency is 455 kHz, the converter will be 455 kHz above or below the tuned frequency. That can mix with other signals to produce an image on the tuned frequency. **A receiver tuned to 14.300 MHz with a 455 kHz IF frequency would also receive a signal at 15.210 MHz.** The local oscillator will be running at 14.755 MHz to generate a 455 kHz IF. But a signal at 15.210 would also mix with 14.755 resulting in 455 kHz out. *Hint: A shortcut is to double the IF and add it to the tuned frequency.*

Selectivity defines how narrow a filter is. **If the filter is too wide, undesired signals may be heard.**

An amateur RTTY filter should be about 300 Hz wide.

An amateur SSB filter would be 2.4 kHz wide. *Hint: think about how wide the mode is. The other answers are way off.*

A roofing filter is applied before the more sensitive IF circuits. **A narrow-band roofing filter effects receiver performance because it improves the dynamic range by attenuating strong signals near the receive frequency.** *Hint: Filters attenuate signals near the receive frequency.*

MORE RECEIVER PERFORMANCE CHARACTERISTICS (E4D)

The blocking dynamic range of a receiver is the difference in dB between the noise floor and the level of an incoming signal that will cause 1 dB of gain compression. *Cheat: Recognize the answer with 1 dB.*

The term for a reduction in receiver sensitivity caused by a strong signal near the received frequency is desensitization. *Hint: Reduction in sensitivity is desensitization.*
Desensitization can be caused by strong adjacent channel signals.

A way to reduce desensitization is to decrease the RF bandwidth of the receiver. *Hint: Switch in a narrower filter to cut down the offending signal.*

Two problems caused by poor dynamic range in a receiver are cross modulation of the desired signal and desensitization from strong adjacent signals. *Hint: Strong adjacent signals desensitize a receiver with poor dynamic range.*

Unwanted signals generated by the mixing of two or more signals is called intermodulation interference.

Intermodulation interference can occur between two repeaters when the repeaters are in close proximity and the signals mix in the final amplifier of one or both. *Hint: Modulation is mixing.*

To reduce or eliminate intermodulation interference in a repeater use a properly terminated circulator at the output of the transmitter. *Hint: To eliminate, terminate.* A circulator is a one-way valve that shunts off signals coming down the transmission line and sends them to a dummy load (properly terminated).

If a receiver tuned to 146.70 MHz, and a nearby station transmits on 146.52 MHz, the transmitter frequencies would produce an intermodulation-product signal in the receiver are 146.34 MHz and 146.61 MHz. *Hint/Cheat: Take the average of the two (146.70 + 146.52)/2 and look for the answer that has 146.61. You'll only solve one but that is enough to choose the correct answer.*

The most significant effect of an off-frequency signal when it is causing cross-modulation interference to a desired signal is the off-frequency signal is heard in addition to the desired signal. *Hint: isn't that the very definition of interference?*

Intermodulation in electronic circuits is caused by non-linear circuits or devices.
The purpose of a preselector in a receiver is to increase rejection of unwanted signals. *Hint: It 'preselects signals.*

A third-order intercept level of 40 dBm means a pair of 40 dBm signals will generate a third-order intermodulation product with the same level as the input signals. *Cheat: A pair are the same.*

Third-order intermodulation products are of particular interest because the third order product of two signals which are in a band of interest will also likely be within the band. *Hint: You are interested because the products will be in the band you are working.*

NOISE SUPPRESSION (E4E)

A type of noise that can be reduced by a receiver noise blanker is ignition noise. Noise blankers work best on repetitive noise like the pops from spark plugs.

A receiver noise blanker might be able to remove signals which appear across a wide bandwidth.

Conducted and radiated noise from an automobile alternator can be suppressed by connecting the radio's power leads directly to the battery and by installing coaxial capacitors in line with the alternator leads. *Hint: Connecting directly to the battery is the best way to power a mobile rig. Look for that in the answer.*

DSP is digital signal processing. Like an SDR, the signal is digitized, and computer power applied to form filters and accomplish noise reduction.
Receiver noise best reduced with a DSP noise filter would be
Broadband white noise.
Ignition noise
Power line noise.
All of these choices are correct.
Hint: DSP is a powerful interference fighting tool, so all the choices are correct.

Noise from an electric motor can be suppressed by installing a brute-force AC-line filter in series with the motor leads. *Hint: Suppress noise with brute-force.*

A major cause of atmospheric static is thunderstorms. You can hear the static over great distances and even if the storm is far from you.

To determine if noise is being generated in your home, turn off the AC power at the main circuit breaker and listen on a battery-powered radio.

The signal picked up by electrical wiring near a radio antenna is common-mode signal at the frequency of the transmitter. *Hint: If the signal is picked up from an antenna, it has to be at the transmitter frequency.*

An undesirable effect when using an IF noise blanker is signals may appear excessively wide even if they meet emission standards. *Cheat: The IF (intermediate frequency) is at radio frequencies. The other answers all relate to audio. If you can't recognize the correct answer, read the question and answers carefully for clues.*

A common characteristic of interference caused by touch controlled electrical devices is:
The interfering signal sounds like hum.
The interfering signal may drift slowly.
The signal may be several kHz wide and repeat at regular intervals.
All these choices are correct.
Hint: Touch controlled devices are nasty RF generators, so all the choices are correct.

If you hear combinations of local AM broadcast stations within MF or HF ham bands, it most likely is caused by nearby corroded metal joints mixing and re-radiating the broadcast signals. Corroded metal joints can act as diodes and mix signals causing intermodulation.

One disadvantage of automatic DSP notch-filtering on CW signals is it removes the desired signal at the same time. The automatic filter removes all tones including the CW tone.

AMATEUR PRACTICES

Loud roaring or buzzing that comes and goes could be:
Arcing contacts in a thermostatically controlled device.
Defective doorbell or doorbell transformer.
Malfunctioning illuminated advertising display.
All of these choices are correct.
Hint: All operate intermittently are therefore cause noise that comes and goes.

A nearby computer might cause unstable modulated or unmodulated signals at specific frequencies. *Hint: Computer circuits can generate "birdies" – signals heard at specific frequencies.*

Shielded cables can radiate or receive interference due to common mode currents on the shield and conductors. *Hint: It is common to the shield and conductors.* Solve that by putting ferrite chokes on the cable.

Current that flows equally on all conductors is called common-mode current. *Hint: It is common to all the conductors.*

ELECTRICAL PRINCIPLES (E5)

RESONANCE AND Q (E5A)

The voltage across reactance in series can be larger than the voltage applied due to resonance. Those high voltages can cause of arcing in your antenna tuner. *Hint: Think of resonance as multiplying.*

Resonance in an electrical circuit is the frequency at which the capacitive reactance equals the inductive reactance. When the two are equal, they cancel each other, and the circuit resonates.

The magnitude of the impedance in a series RLC circuit at resonance is approximately equal to the circuit resistance. *Hint: "RLC" refers to resister, inductor, and capacitor. If the inductor and capacitor cancel each other out, all that is left is resistance.*

The magnitude of impedance with a resistor, inductor and capacitor all in parallel at resonance is equal to the circuit resistance. *Hint: In series, or in parallel, at resonance, the answer is the same as above.*

The magnitude of the current in a series RLC circuit at resonance is maximum. *Hint: The magnitude of the current is maximum because the inductor and capacitor have canceled each other out.*

The magnitude of the circulating current in a parallel LC circuit at resonance is maximum. *Hint: The <u>circulating</u> current in the circuit is maximum. The inductive and capacitive reactance are in parallel and the current gets transferred back and forth between them.*

The magnitude of the current at the input of a parallel RLC circuit at resonance is minimum. *Hint: The circulating current stays in the circuit and doesn't draw current from the supply. The glass is full. The input current is at a minimum.*

The phase relationship between the voltage and current in a series resonate circuit is the voltage and current are in phase. *Hint: In a series circuit, the inductance and capacitance have canceled each other and canceled any phase changes each created.*

"Q" stands for reactive quotient and is a measure of selectivity. A high Q circuit is highly selective and narrow banded. **The effect of increasing the Q in an impedance matching network is the matching bandwidth is decreased.**

The Q of an RLC parallel resonate circuit is calculated by resistance divided by reactance. The Q of an RLC series resonate circuit is calculated by the reactance divided by the resistance. *Cheat: Parallel resistance, series reactance.*

Increasing Q in a resonant circuit increases internal voltages and circulating currents. *Hint: Increasing Q increases.*

To increase Q in inductors and capacitors, lower losses. *Hint: The lossy resistance is what makes the circuit less sharp.*

The half-power bandwidth of a parallel resonate circuit that has a resonant frequency of 3.7 MHz, and a Q of 118 is 31.4 kHz.
The half-power bandwidth of a parallel resonate circuit that has a resonate frequency of 7.1 MHz, and a Q of 150 is 47.2 kHz.

Solution: The bandwidth is frequency divided by Q. 3.7 MHz/118 = .3135 MHz is 31.35 kHz and 31.4 kHz is the closest answer. The answers are far enough apart you can get sloppy with your decimal points. Look for the answer with the correct integers.

The resonate frequency of a series RLC circuit if R is 22 ohms, L is 50 microhenrys and C is 40 picofarads is 3.56 MHz.
The resonate frequency of a parallel RLC circuit if the R is 32 ohms, L is 50 microhenrys and C is 10 picofards is 7.12 MHz.
Resonant frequency is determined by the formula F=1/(2∏ X √LC). If that math is too much for you, as it is for me, there are online calculators so I don't feel bad about giving you this cheat: *The answer is always in the 80 or 40-meter ham band.*

TIME AND PHASE (E5B)

The term for the time required for the capacitor in an RC circuit to be charged to 63.2% of the applied voltage is "one time constant."
The term for the time it takes a charged capacitor to be discharged to 36.8% of its initial voltage is "one time constant."

The time constant of a circuit having two 220 microfarad capacitors and two 1 megaohm resistors in parallel is 220 seconds. Solve: The formula is resistance times capacitance. Two 220 microfarad capacitor in parallel are 440 microfarads. Two 1 megaohm resistors in parallel are .5 megaohms. 440 x .5 = 220. *Cheat: you could just remember the answer is the value of the capacitor.*

Susceptance is the reverse of reactance.

The letter B is used to represent susceptance.

When the phase angle of a reactance converted to susceptance the sign is reversed. *Hint: Reactance and susceptance are opposite ways of expressing the same thing – resistance to AC. Phase angle has a sign.*

The magnitude of the reactance when it is converted to susceptance is the reciprocal of the magnitude of the reactance. *Hint: Magnitude doesn't have a sign so the opposite is a reciprocal.*

Impedance causes the voltage and current to get out of phase. In an inductive circuit, voltage leads the current. In a capacitive circuit, voltage lags the current. Remember "ELI the ICEman." E is voltage, L is inductive, I is current, C is capacitive.

The phase angle between the voltage across and current through a series RLC circuit if XC is 500 ohms, R is 1 kilohm and XL is 250 ohms is 14 degrees with the voltage lagging the current. *Hint: We know the voltage is lagging the current because the XC is greater than the XL. That eliminates 2 answers. The rest of the math is staggering. Cheat: The answer to all three questions is 14. Look for the greater of XL or XC and ask ELI to tell you which is leading and lagging.*

The phase angled between the voltage across and the current through a series RLC circuit if the XC is 100 ohms, the R is 100 ohms and XL is 75 ohms is 14 degrees with the voltage lagging the current. *Hint: Same as above.*

The phase angle between the voltage across and current through a series RLC circuit if the XC is 25 ohms , R is 100 ohms and XL is 50 ohms *Hint: Here the circuit is more inductive to voltage leads current. The rest of the answer is 14 again.*

The relationship between current through a capacitor and the voltage across it are that the current leads the voltage by 90 degrees. *Hint: Current leads voltage in a capacitor. ELI the ICEman told me so*

The relationship between the current through and the voltage across an inductor is voltage leads the current by 90 degrees. *Hint: Voltage leads current in an inductor says ELI.*

Admittance is the term that is the inverse of impedance. *Hint: Impedance impedes and admittance admits.*

COORDINATE SYTEMS AND PHASORS (E5C)

A capacitive reactance in rectangular notation is −jx. *Hint: Look at ELI the ICEman's voltage. This is a capacitive circuit, so the voltage is behind (-j)*

An impedance of 50-j25 represents 50 ohms resistance with 25 ohms capacitive reactance. The first number is ohms resistance. The second is reactance. Since the number is negative, we know the voltage is lagging, and the circuit must be capacitive.

Impedances described in polar coordinates are a phase angle and amplitude.

A vector is a quantity with both magnitude and an angular component. A vector is an arrow. The length is the magnitude and the direction it points is the angle.

In polar coordinates, an inductive reactance has a positive phase angle. Hint: *This is an inductive circuit so the voltage is ahead of current (positive angle).*

ELECTRICAL PRINCIPLES

In polar coordinates a capacitive reactance has a negative phase angle. *Hint: Capacitive reactance puts the voltage behind the current (negative angle).*

The diagram used to show the phase relationship between impedances at a given frequency is called a phasor. *Cheat: If you see a Star Trek weapon in the answer, it is correct.*

The coordinate system used to display resistive, inductive and/or capacitive reactance components of impedance is rectangular coordinates. Figure E5-2 is an example.

When using rectangular coordinates to graph the impedance of a circuit, the horizontal axis is the resistive component. *Cheat: Horizontal sounds like reziztive.*

If the impedance falls to the right side of the graph on the horizontal axis, the circuit is equivalent to a pure resistance.

The vertical axis represent the reactive component. *Cheat: Vertical and reactive both have vees.*

The two numbers used to define a point on a graph are the coordinate values on the horizontal and vertical axis.

The following questions refer to figure E5-2.
The point representing a 400-ohm resistor and a 38 picofarad capacitor at 14 MHz is point 4. Solve: We know the answer lies on the 400-ohm line of the horizontal axis (resistance). We know the circuit is capacitive so it will have a minus sign and be on the lower quadrant on the vertical axis. Point 4 it is by default!

ELECTRICAL PRINCIPLES

Figure E5-2

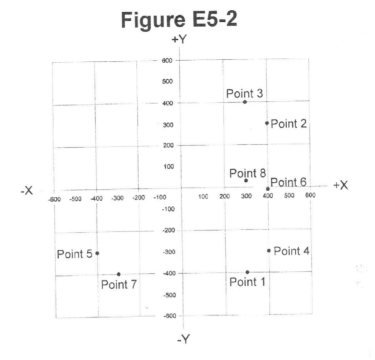

The point representing a circuit has a 300-ohm resistor, and an 18 millihenry inductor at 3.5 MHz is point 3. Solve: Reactance is 2π times frequency times inductance. 2x3.14x18x3.5= 395. We know the circuit is inductive to we are looking to the top following the 300-ohm horizontal line. Point 3 is the correct answer, where 300 and 395 cross.

The point representing a circuit that has a 300-ohm resistor and a 19 picofarad capacitor at 21.200 MHz is point 1. Solve: We are on the 300-ohm line of the horizontal axis, and the circuit is capacitive, so we must be in the lower quadrant. Point1 it is by default.

The point representing a circuit with a 300-ohm resistor, a .64 microhenry inductor and an 85 picofarad capacitor at 24.9 MHz is point 8. Solve: We are on the 300-ohm line, and the circuit is slightly inductive. Point 8 by default.

Extra Class – The Easy Way   Page 55

AC AND RF ENERGY (E5D)

The result of skin effect is as frequency increases; RF flows in a thinner layer of the conductor, closer to the surface.

It is important to keep lead lengths short for components in circuits for VHF and above because you want to avoid unwanted inductive reactance.

Short connections are necessary at microwave frequencies to reduce phase shift along the connection. *Hint: This is another way of saying you don't want any inductive reactance.*

The parasitic characteristic that increases with conductor length is inductance. *Hint: Longer wire, more inductance.*

Microstrip is precision printed circuit conductors above a ground plane that provide a constant impedance at microwave frequencies. *Hint: Way too much information! Making a microstrip requires precision. It is a precision printed circuit.*

The direction of a magnetic field oriented around a conductor is in a direction determined by the left-hand rule. Put your left hand around the wire with your thumb pointed in the direction of the current. The magnetic field is in the direction of your fingers.

The strength of the magnetic field around a conductor is determined by the amount of current flowing. More current, more magnetic field.

The energy stored in an electromagnetic or electrostatic field is potential energy. *Hint: It is stored energy so it is potential.*

The reactive power in an AC circuit that has ideal inductors and capacitors is repeatedly exchanged between the associated magnetic and electric fields but is not dissipated. *Hint: If the components are ideal, energy is not dissipated. Look for "not dissipated" in the correct answer.*

Reactive power is wattless and non-productive power.

The true power in an AC circuit where the voltage and current are out of phase can be determined by multiplying the apparent power times the power factor. When the current and voltage are out of phase, some power is lost so the apparent power is lower. Multiply apparent power by the power factor to see what went into the circuit.

The power factor of an R-L circuit having a 60-degree phase angle between the voltage and the current is .5 Solve: The answer is the cosine of the phase angle. The cosine of 60 degrees is .5. You will have to bring a scientific calculator to the test.
The power factor of an R-L circuit with a 30-degree phase angle is .866. Solve. The cosine of 30 degrees is .866.
The power factor of an R0L circuit with a 45-degree phase angle is .707. Solve: the cosine of 45 degrees is .707.

If a circuit has a power factor of .2 and the input is 100VAC at 4 amperes, 80 watts are consumed. The total watts are 400 and .2 of that is consumed. Solve: 100 x 4 = 400 x .2 = 80.

If a circuit has a power factor of .6 and the input is 200VAC at 5 amperes, 600 watts are consumed. Solve: 200 x 5 x .6 = 600.

ELECTRICAL PRINCIPLES

If the apparent power is 500VA (watts) and the power factor is .71, the power consumed is 355 watts. Solve: 500 x .71 = 355.

The power consumed in a circuit consisting of a 100-ohm resistor in series with a 100-ohm inductive reactance drawing 1 ampere is 100 watts. *Foul! This is a trick question. It assumes no loss in the inductor. How much power is consumed by the resistor portion of the circuit? You are supposed to ignore the power consumed in the inductor. The answer is $P = I^2 R$. I^2 is 1 times the R (100) is 100 watts.*

CIRCUIT COMPONENTS (E6)

SEMICONDUCTORS (E6A)

Gallium arsenide is used as a semiconductor material in preference to germanium or silicon in microwave circuits.

The semiconductor material that contains excess free electrons is N-type. *Hint: Excess free electrons would be **N**egatively charged.*

A PN-junction does not conduct current when reverse biased because holes in the P-type material and electrons in the N-type material are separated by the applied voltage widening the depletion region. *Hint: It is enough to remember the materials are separated by the voltage.*

The name given to an impurity that adds holes to a semiconductor crystal structure is acceptor impurity. *Hint: Holes accept electrons.*

The alpha of a bipolar transistor is the change in collector current with respect to emitter current. *Hint: For those who remember tubes, the emitter is like the grid.*

The beta of a bipolar transistor is the change in collector current with regard to base current. The change between the collector and base is gain. *Cheat: Beta and base current.*

A silicon NPN transistor is biased on when the base-to-emitter voltage is approximately .6 - .7 volts. *Hint: It is biased by voltage, eliminating 2 of the answers. The bias voltage is small, eliminating the other answer.*

The term indicating the frequency at which a grounded-base current gain of a transistor has decreased to .7 of the gain obtainable at 1 kHz is called the alpha cutoff frequency.

A depletion-mode FET is an FET that exhibits a current flow between source and drain with no gate voltage is applied. *Hint Current flow with no gate would deplete.*

Figure E6-2

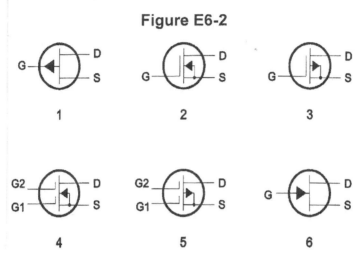

In Figure E6-2 the symbol for an N-channel dual-gate MOSFET is number 4. *Hint: It is dual gate and only two of the figures show 2 gates. It is N channel so the arrow is pointed in. Note: This is the opposite of bipolar transistors where the N type has the arrow not pointed in.*

The symbol for a P-channel junction FET is number 1. *Hint: A P-channel FET has the arrow pointed out. These are the only two symbols you need to know from Figure E6-2.*

Many MOSFET devices have internally connected Zener diodes on the gates to reduce the chance of the gate insulation being punctured by static discharges or excessive voltages.

CMOS means Complementary Metal Oxide Semiconductor. *Hint: It has to be metal because mica doesn't conduct.*

The DC input impedance at the gate of a field-effect transistor is high compared to a bipolar transistor which is low impedance.

The semiconductor material with excess holes in the outer shell of electrons is P type. *Hint: Excess holes means no free electrons and a Positive charge.*

The majority of charge carriers in an N-type semiconductor are free electrons. *Hint: electrons carry a charge.*

The three terminals of a field-effect transistor are gate, drain and source.

DIODES (E6B)

The most useful characteristic of a Zener diode is constant voltage drop under conditions of varying current.

The important characteristic of a Schottky diode is less forward voltage drop.
Cheat: Look for "voltage drop" in both answers.

The special type of diode capable of both amplification and oscillation is a tunnel diode.
Hint: You can echo (oscillate) in a tunnel.

The semiconductor device designed for use as a voltage-controlled capacitor is a varactor diode.
Hint: The voltage varies the capacitance.
The characteristic of a PIN diode that makes it useful as an RF switch or attenuator is a large region of intrinsic material. PIN means the layers

are P Intrinsic and N. *Hint: "Intrinsic" implies natural. It has a region where no impurities are added to create holes and free electrons.*

To control the attenuation of RF signals by a PIN diode, use forward DC bias current. *Hint: Give the signals a boost with forward bias.*

A common use for a PIN diode an RF switch. A transmit/receive switch uses a PIN diode.

A common use of a hot-carrier diode is as a VHF/UHF mixer or detector.

When a junction diode fails due to excessive current, the mechanism is excessive junction temperature. *Hint: Excessive current causes excessive heat which causes a failure.*

A type of semiconductor diode is a metal-semiconductor junction. *Hint: The question and answer both contain "semiconductor."*

A common use for a point contact diode is as an RF detector. *Hint: An RF detector helps you make a contact.*

See the next page for Figure E6-3.
In Figure E6-3 the symbol for a light-emitting diode is number 5. *Hint: The arrows are rays of light.*

The bias to light an LED is forward bias. *Hint: It is conducting – moving forward. This is the only symbol you need to know on Figure E6-3.*

Figure E6-3

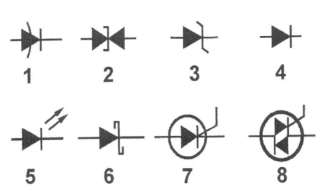

DIGITAL ICS (E6C)

A comparator is a device that compares two voltages or currents and outputs a digital signal indicating which is larger. **When the level of a comparator's input signal crosses the threshold, the comparator changes its output state.**

The function of hysteresis in a comparator is to prevent input noise from causing unstable output signals.

Tri-state logic is logic devices with 0,1 and high impedance output states. *Hint: Three output states.*

The primary advantage of tri-state logic is the ability to connect many device outputs to a common bus.

The advantage of CMOS logic devices over TTL devices is lower power consumption.

CMOS digital integrated circuits have high immunity to noise on the input signal or power supply because the input switching threshold is about one-half the power supply voltage. *Hint:*

CIRCUIT COMPONENTS

It switches long before power supply voltage variations kick in.

A pull-up or pull-down resistor is connected to the positive or negative supply line used to establish a voltage when an input or output is an open circuit. *Hint: Pull up or pull down means it could be positive or negative.*

A Programmable Logic Device is a programmable collection of logic gates and circuits in a single integrated circuit.

Figure E6-5

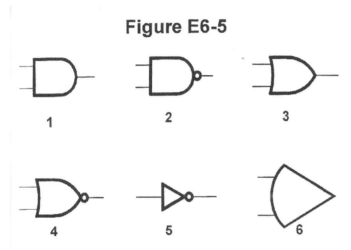

Figure E6-5 has symbols for logic circuits.

Hint: The little circle means the output is reversed. It is a N. **In figure E6-5 the symbol for a NAND gate is number 2.** *Hint: The two inputs are on a straight line. It is an AND. See page 74.*

The symbol for a NOR gate is number 4. *Hint: The inputs are on a curved line. OR.*

The symbol for the NOT operation (inverter) is number 5. *Hint: One input and one output is all that is required to invert.*

BiCMOS logic is and integrated circuit logic family using both bipolar and CMOS transistors. *Hint: Bi(polar) and CMOS.*

An advantage of BiCMOS logic is it has the high input impedance of CMOS and the low output impedance of bipolar transistors. *Hint: High input impedance is a good thing because it doesn't load down the circuit.*

The advantage of a Programmable Gate Array in a logic circuit is that complex logic functions can be created in a single integrated chip. *Hint: An array is a series of logic functions.*

TOROIDS (E6D)

The number of turns it will take to produce a 5-microhenry inductor using a powdered iron toroidal core that has an induction index of 40 microhenrys/100 turns is 35. Solve: Take the desired inductance and divide it by the turns ratio. Take the square root of that. 5/40 = .125. The square root of .125 is .353. Since the index was for 100 turns, we multiply .353 by 100 to get 35.3 turns.

The number of turns required to produce a 1-mh inductor using a core that has an inductance index value of 523 millihenrys/1000 turns is 43. Solve: 1/523 = .0019 and the square root is .0437. The index was for 1000 turns, so the number of turns is 43.

The equivalent circuit of a quartz crystal is motional capacitance, motional inductance and loss resistance in series, all in parallel with a shunt capacitor representing electrode and stray capacitance. *Cheat: Horrors! You have got to be kidding. Just look for the answer with "shunt" in it and bypass all this.*

CIRCUIT COMPONENTS

An aspect of the piezoelectric effect is mechanical deformation of material by the application of a voltage. *Hint: We usually think of it the other way around, voltage generated by pressing the material like a grill lighter.*

The materials commonly used as a slug core in a variable inductor are ferrite and brass.

A reason to use ferrite cores instead of powdered-iron is ferrite require fewer turns to produce a given inductance value.

A reason to use powdered-iron instead of ferrite it the powdered-iron maintain their characteristics at higher currents. *Hint: Iron can handle higher current because iron can withstand heat better.*

The material property that determines the inductance of a toroidal inductor is its permeability.

The usable frequency range of toroidal cores, assuming a correct selection of the core material, is from less than 20 hz to 300 MHz. Toroids work up through VHF. Above that, the capacitance between the windings becomes too great and it starts to act more like a capacitor than an inductor.

The primary cause of inductor self-resonance is inter-turn capacitance.

The devices commonly used as VHF and UFH parasitic suppressors at the input and output terminals of a transistor HF amplifier are ferrite beads. *Hint: Ferrite beads are toroids a wire goes through instead of wrapping around. The beads act as chokes.*

The primary advantage of using a toroidal core over a solenoidal core is the toroidal core confines most of the magnetic field within the core material. A solenoidal core is a bar of ferrite with the wire wrapped around. The circular toroid keeps the magnetic field contained.

Saturation of a ferrite core inductor is when the ability of the core to store magnetic energy has been exceeded. *Hint: The core is saturated.*

Core saturation should be avoided because harmonics and distortion could result. *Hint: The core can't store any more magnetic energy, so there is distortion.*

The material that decreases inductance when inserted into a coil is brass. *Hint: Brass is not magnetic.*

The current in the primary winding of a transformer with no load attached is called magnetizing current. *Hint: It is just enough to magnetize the core but no more.*

The common name of a capacitor connected across a transformer secondary that is used to absorb transient voltage spikes is a snubber capacitor. *Hint: It snubs out the spikes.*

ANALOG ICS (E6E)

A charge-coupled device (CCD) samples an analog signal and passes it in stages from the input to the output.

An example of a through-hole device is a DIP. Dual Inline Package is a type of computer chip. It has leads that are soldered through holes in the circuit board.

A characteristic of DIP packaging is two rows of connecting pins placed on opposite sides of the package. *Hint: Dual Inline Package.*

MMIC means monolithic Microwave Integrated Circuit. **The material likely to provide the highest frequency of operation when used in MMICS is gallium nitride.** *Hint: Gallium is used at high frequencies. Gallium is also the answer in another question.*

The most common input and output impedance of a circuit that use MMICs is 50 ohms. *Hint: Just like our coax.*

The characteristic of MMIC that makes it popular for VHF through microwave circuits is controlled gain, low noise figure, and constant input and output impedance over the specified frequency range. *Hint: You don't need to remember all of that. Just pick "low noise."*

The noise figure typical for a low-noise UHF preamplifier is 2 dB. *Hint: The pre-amplifier will add some noise but not a lot. The other answers are either negative or too high.*

An MMIC-based microwave amplifier uses microstrip construction. Hint: Microwave = microstrip.

The voltage from a power supply normally furnished to the monolithic microwave integrated circuit (MMIC) is through a resistor and RF choke connected to the amplifier output lead. Hint: The power supply attaches to the output just like every other amplifier.

The component package most suitable for use above the HF range are surface mount. Hint: Surface mount has the shortest leads and short leads are required on high frequencies.

The packaging technique in which leaderless components are soldered directly to the circuit board is called surface mount.

High power RF amplifier ICs and transistors are sometimes mounted in ceramic packages for better dissipation of heat. Hint: Ceramic can withstand and dissipate heat.

OPTICAL COMPONENTS (E6F)

Photoconductivity is increased conductivity of an illuminated semiconductor.
When light shines on a photoconductive material the conductivity increases.

A material most affected by photoconducivity is a crystalline semiconductor.

The photovoltaic effect is the conversion of light to electrical energy.

The efficiency of a photovoltaic cell is the relative fraction of light that is converted to

Extra Class – The Easy Way   Page 69

CIRCUIT COMPONENTS

current. *Hint: The more efficient, the more current it produces for a given amount of light.*

The most common type of photovoltaic cell used for electrical power is silicon. *Hint: Solar panels are made of silicon.*

The approximate voltage produced by a fully illuminated silicon photovoltaic cell is 0.5 V. Each cell produces 0.5 volts, and the cells are connected in series to output a higher voltage.

The energy from light falling on a photovoltaic cell is aborbed by electrons. *Hint: Electrons are what make electricity.*

The most common configuration of an optoisolator or optocoupler is an LED and a phototransistor. Two circuits are isolated (not electrically connected) but signals pass by light.

An optoisolator is often used in conjuction with solid state circuits when switching 120VAC because optoisolators provide a high degree of electrical isolation between the control circuit and the circuit being switched. *Hint: Just remember optoisolators provide isolation and look for that word in the answer.*

An optical shaft encoder detects rotation of a control by interrupting a light source with a patterned wheel. *Hint: It senses light pulses when the shaft turns.*

A solid state relay is a device that uses semiconductors to implement the functions of an electromechanical relay. *Hint: A solid state relay is a relay made with solid state devices (semiconductors).*

A liquid crystal display (LCD) is a display utilizing a crystalline liquid and polarizing filter which becomes opaque when voltage is applied. *Hint: Look for the answer that repeats the question. The display is liquid crystal, so it uses a crystalline liquid.*

Liquid crystal displays are hard to view through polarized lenses. *Hint: The display uses a polarizing filter.* You have a hard time reading an LCD when wearing polarized sunglasses.

PRACTICAL CIRCUITS (E7)

DIGITAL CIRCUITS (E7A)

A bi-stable circuit is a flip-flop. *Hint: "Bi" means it has two states, flip and flop.*

The circuit that can divide the frequency of a pulse train by 2 is a flip-flop.
To divide a signal frequency by 4 requires 2 flip-flops.

The function of a decade counter digital IC is to produce one output signal for every ten input pulses. *Hint: It counts decades or tens.*

A circuit that continuously alternates between two states without an external clock is an astable multivibrator. *Hint: Astable means it alternates continuously. It is never stable.*

A monostable multivibrator is that it switches momentarily to the opposite binary state and then returns to its original state after a set time. *Hint: Monostable means it returns to one state.*

A NAND gate produces a logic 0 when all inputs are logic 1. *Hint: N means it produces an opposite logic and AND mean both inputs are the same.*

An OR gate produces a logic 1 if any or all inputs are logic 1. *Hint: There is no N in front, so output equals input. OR means either input could be 1.*

A NOR gate produces a logic 0 if any outputs are logic 1. *Hint: The N means it produces an opposite logic. The OR means either input could be 1.*

A truth table is a list of inputs and corresponding outputs for a digital device. *Hint: It is a table showing the various outcomes.*

The logic that defines "1" as a high voltage is positive logic. *Hint: 1 is a positive number.*

The type of logic that defines "0" as a high voltage is negative logic. *Hint: The opposite.*

AMPLIFIERS (E7B)

Amplifiers are classed by how much of the 360-degree signal cycle they operate.

A Class AB amplifier operates more than 180 degrees but less than 360 degrees.

A Class D amplifier uses switching technology to achieve high efficiency.
Switching amplifiers are more efficient than linear amplifiers because the power transistor is at saturation or cut off most of the time, resulting in low power dissipation. Class D only amplifies during a small part of the cycle.

The components of a class D amplifier include a low pass filter to remove switching signal components. *Hint: Switching circuits can generate spurious signals, and the low pass filter knocks them down.*

A Class A amplifier would normally be biased half-way between saturation and cutoff. Class A runs all through the 360-degree cycle so it would have to be biased right in the middle.

To prevent unwanted oscillations in an RF power amplifier, install parasite suppressors and/or

PRACTICAL CIRCUITS

neutralize the stage. Neutralization is introducing some negative feedback to cancel the oscillations. **An RF power amplifier can be neutralized by feeding in a 180-degree out-of-phase portion of the output back to the input.**

An amplifier that reduces or eliminates even harmonics is a push-pull design.

If a Class C amplifier is used to amplify SSB, you would experience signal distortion and excessive feedback. Class C operates less than 50% of the time. Class C is not linear and is suited for CW but not SSB.

When tuning a vacuum tube amplifier that uses a Pi-network, adjust the tuning capacitor for minimum plate current and the loading capacitor for maximum permissible plate current. *Hint: Tune for minimum and load for maximum*

Figure E7-1

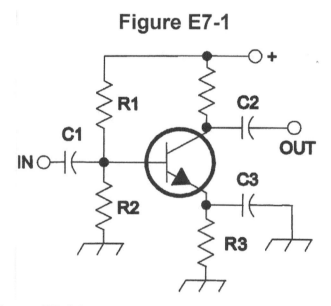

Figure E7-1 is a common emitter transistor amplifier.

There are three questions related to figure E7-1 and you only need to know the purpose of R1/R2 and R3. **In Figure E7-1 the purpose of R1 and R2 is fixed bias.** They set the voltage on the base of the transistor. They form a voltage divider between the + and ground. The voltage on the base stays the same even if the signal changes.

The purpose of R3 is self bias. It is biasing the whole transistor. Amplifiers operate using signal inputs which vary from positive to negative. Biasing establishes the correct operating point of the transistor and, if done correctly, reduces distortion.

The type of amplifier in figure E7-1 is a common emitter.

Figure E7-2

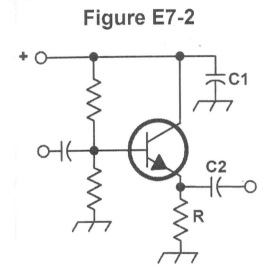

The purpose of R in Figure E7-2 is to be the emitter load. *Hint: See how it is connected to the emitter. This is the only question on this figure.*

To prevent thermal runaway in a bipolar transistor, use a resistor in series with the emitter. That restricts the amount of current that can flow and overheat the transistor.

The effect of intermodulation products in a linear power amplifier are transmission of spurious signals. *Hint: Intermodulation is distortion and can lead to spurious signals.*

Odd-order rather than even-order intermodulation distortion products are of greater concern in linear power amplifiers because they are relatively close in frequency to the desired signal. *Hint: Being close in frequency is a problem.*

A characteristic of a grounded-grid amplifier is low input impedance. *Hint: If the grid is grounded, it is at low impedance.*

FILTERS AND MATCHING NETWORKS (E7C)

Capacitors and inductors in a low-pass filter Pi-network are arranged with a capacitor connected between the input and ground, another capacitor between the output and ground and an inductor is connected between the input and output. *Hint: Another overly complicated answer. The filter is to pass low frequencies so it must shunt off high frequencies. The way to do that is with capacitors to ground from both the input and output. Look for that as part of the answer.*

A property of a T-network with series capacitors and a parallel shunt inductor is a high-pass filter. *Hint: Series capacitors pass high frequencies so it is a high pass filter.*

A pi-L network has an advantage over a regular Pi-network because it offers greater harmonic suppression. *Hint: The extra component (L) offers more suppression.*

A Pi-L network used for matching a vacuum tube final amplifier to a 50-ohm unbalanced output is

a Pi network with an additional series inductor on the output. *Hint: Pi-L adds an L (inductor) to a Pi network.*

An impedance matching circuit transforms a complex impedance to a resistive impedance by canceling the reactive part of the impedance and changing the resistive part to a desired value. *Hint: Impedance matching changes to a desired value.*

The filter with a ripple in the passband and a sharp cutoff is a Chebyshev filter. *Cheat: Recognize the Russian name and if you see it, you have the answer.*

The distinguishing features of an elliptical filter are extremely sharp cutoff with one or more notches in the stop band. *Cheat: "Sharp cutoff" is good in a filter and the answer to both questions.*

To attenuate a carrier signal while receiving a SSB transmission you would use a notch filter. *Hint: It "notches" out the offending carrier.*

The factor having the greatest effect on determining the response shape of a crystal ladder filter is the relative frequencies of the individual crystals. *Hint: The crystal frequencies determine the shape of the filter.*

A crystal lattice filter is a filter with narrow bandwidth and steep skirts made using quartz crystals. *Hint: Several crystals are connected in a lattice making the filter narrower and sharper.*

A Jones filter is a variable bandwidth crystal lattice filter. Normally a crystal filter is fixed-width, but Jones figured out a way to connect crystals and vary the width.

The best choice of a filter for a 2-meter repeater would be a cavity filter.

The common name for a filter network which is the equivalent of two L-networks connected back-to-back with the two inductors in series and capacitors in shunt with the input and output is a Pi. *Hint: The components look like the Greek letter*

An advantage of a Pi-matching network over an L-matching network consisting of a single inductor and capacitor is the Q of Pi networks can be varied depending on the component values chosen. *Hint: Q is determined by the relationship of input resistance to output resistance. A single inductor and capacitor means you can't change both the input and output resistance. Your choices are limited.*

The mode most affected by non-linear phase response in a receiver IF filter is digital. *Hint: Digital modes require decoding and need good data.*

POWER SUPPLIES (E7D)

One characteristic of a linear electronic voltage regulator is the conduction of a control element is varied to maintain a constant output. *Hint: It is a regulator so it regulates to "maintain a constant output."*

One characteristic of a switching electronic voltage regulator is the duty cycle is changed to produce a constant average output voltage. *Hint: Switching is changing the duty cycle.*

The device typically used as a stable reference voltage in a linear voltage regulator is a Zener diode. Zener diodes are voltage regulators.

The linear voltage regulator that makes the most efficient use of the primary power source is a series regulator. *Hint: "Regulator" in the question and the answer. Because it is in series, all the power passes to the load. A shunt would dump some to ground.*

The linear voltage regulator that places a constant load on the unregulated voltage source is a shunt regulator. *Hint: "Regulator" in the question and answer. The shunt keeps a constant load dumping excess to ground.*

Figure E7- 3

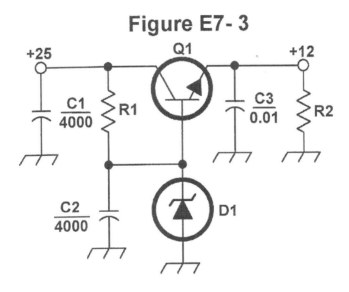

There are only three possible questions related to Figure E7-3. **The circuit shown in Figure E7-3 is a linear voltage regulator.**

The purpose of Q1 in Figure E7-3 is to increase the current-handling capability of the regulator. *Hint: The current is passing through a big power transistor instead of the regulator.*

PRACTICAL CIRCUITS

The purpose of C2 is to bypass hum around D1. *Hint: Capacitors are often used to bypass hum.*

The reason to use a charge controller with a solar power system is to prevent battery damage due to overcharge. *Hint: A charge controller controls the amount of charge.*

The reason a high-frequency switching type high voltage power supply can be both less expensive and lighter in weight than a conventional power supply is the high-frequency inverter design uses much smaller transformers and filter components. *Hint: "High frequency" in both the question and answer.*

The circuit element controlled by a series analog voltage regulator to maintain a constant output voltage is a pass transistor. *Hint: It is a series circuit, so voltage passes through. Q1, in Figure E7-3 above, is a pass transistor.*

The drop-out voltage of an analog voltage regulator is the minimum input-to-output voltage required to maintain regulation. *Hint: Without the minimum input-to-output voltage, it drops out and stops regulating.*

The equation for calculating power dissipation by a series connected linear voltage regulator is voltage difference from input to output multiplied by output current. *Hint: You are solving for dissipation, the power kept in the regulator and that depends on how much of a difference there is between input and output.*

The purpose of a "bleeder" resistor in a conventional unregulated power supply to improve output voltage regulation. *Hint: It provides a more constant load.*

The purpose of a "step-start" circuit in a high-voltage power supply is to allow the filter capacitors to charge gradually. *Hint: It starts in steps to avoid a voltage surge.*

When several electrolytic filter capacitors are connected in series to increase the operating voltage of a power supply filter circuit, resistors should be connected across each capacitor to Equalize, as much as possible the voltage drop across each capacitor.
Provide a safety bleeder to discharge the capacitor when the supply is off.
Provide a minimum load current to reduce voltage excursions at light loads.
All of these choices are correct.
Cheat: A really long question deserves a really long answer.

MODULATING AND DEMODULATING (E7E)

FM phone emissions can be generated by a reactance modulator on the oscillator. *Hint: It is FM, so you modulate the oscillator and a reactance modulator is one way to do that.*

The function of a reactance modulator is to produce FM using electrically variable inductance or capacitance. *Hint: "Reactance modulator," tells you it is FM. You vary frequency with inductors and capacitors.*

An analog phase modulator produces FM signals by varying the tuning of the amplifier tank circuit. *Hint: "Phase modulator," tells you it is FM. You vary frequency by varying tuning.*

One way to generate a single-sideband phone signal is using a balanced modulator followed by a filter. *Hint: You filter off the other sideband.*

PRACTICAL CIRCUITS

The circuit added to an FM transmitter to boost the higher audio frequencies is a pre-emphasis network. *Hint: It emphasizes the higher audio frequencies.*

De-emphasis is used in FM communications receivers for compatibility with transmitters using phase modulation. It undoes the emphasis added by a pre-emphasis circuit and thereby makes the signal compatible.

The term baseband means the frequency components present in the modulating signal. *Hint: The components are the basis of the modulating signal.*

The principal frequencies that appear at the output of a mixer circuit are the two input frequencies along with their sum and difference frequencies. *Hint: It is a mixer circuit so there are two input frequencies, and the result of the mixing is a sum and a difference.*

When an excessive amount of signal energy reaches a mixer circuit, spurious mixer products are generated. *Hint: Overloading a circuit produces spurious products.*

Rectification, detection, and demodulation are used interchangeably. The three mean the same thing when referring to receiver functions.
A diode detector functions by rectification and filtering of RF signals. *Hint: A detector detects RF.*

The detector used for demodulating SSB signals is called a product detector.

A frequency discriminator stage in an FM receiver is a circuit for detecting FM signals.

DSP FILTERING AND SOFTWARE DEFINED RADIO (E7F)

Direct digital conversion in a software defined radio refers to incoming RF being digitized by an analog-to-digital converter without being mixed with a local oscillator circuit. *Hint: Digital conversion digitizes.* It is direct because the RF is not first converted to an IF with a local oscillator but you don't need to know that to answer the question.

The digital signal processing audio filter used to remove unwanted noise from a received SSB signal is an adaptive filter. *Hint: Digital filters use computing power to look for patterns and adapt to eliminate the noise.*

The digital processing filter used to generate an SSB signal is a Hilbert-transform filter.

A common method of generating an SSB signal using digital signal processing is to combine signals with a quadrature phase relationship.

An analog signal must be sampled by an analog-to-digital converter so that signal can be accurately reproduced at twice the rate of the highest frequency component of the signal. To convert from analog to digital requires double the sampling. To convert from digital to analog only requires half the sampling.

The aspect of a receiver analog-to-digital conversion that determines the maximum receive bandwidth of a Direct Digital Conversion SDR is the sample rate. *Hint: How much bandwidth can be converted is determined by the sample rate.*

PRACTICAL CIRCUITS

The minimum number of bits required for an analog-to-digital converter to sample with a range of 1 volt at a resolution of 1 millivolt is 10 bits. Solve: The resolution is 1/1000 of the range. It takes 10 bits to count to a thousand. $2^{10} = 1024$.

The function of a Fast Fourier Transform is to convert digital signals from the time domain to the frequency domain. *Cheat: If it is fast, it must have something to do with time.*

The digital process applied to I and Q signals in order to recover the baseband modulation information is Fast Fourier Transform.

The letters I and Q in I/Q modulation represent In-phase and Quadrature. *Hint: Recognize "quadrature."* Don't confuse it with the Q factor which is a measure of a filter's selectivity.

Decimation with regard to digital filters is reducing the effective sample rate by removing samples. When the Romans conquered, they would often kill every tenth enemy captured, a process called decimation. Decimation reduced the enemy's effectiveness by removing a sample.

An anti-aliasing digital filter is required in a digital decimator to remove high-frequency signal components which would otherwise be reproduced as lower frequency components.

The minimum detectible signal level for an SDR in the absence of atmospheric or thermal noise is set by the reference voltage level and sample width. *Cheat: Can't we just say the signal level is set by voltage?*

The function of taps in a digital processing filter is to provide incremental signal delays for filter

Page 84   Extra Class – The Easy Way

algorithms. The amount of memory devoted to the filter is set by the number of taps. More taps means it uses more memory but develops a better filter.

A digital signal processing filter could create a sharper filter is it had more taps.

The advantage of a Finite Impulse Response (FIR) filter vs. an Infinite Impulse Response (IIR) filter would be FIR filters delay all frequency components of the signal by the same amount. *Hint: The delay is "finite," a set amount which stays the same.*

To adjust the sampling rate of a digital signal by a factor of ¾, interpolate by a factor of three, then decimate by a factor of four. *Hint: Recognize to adjust the sampling rate you "interpolate" and "decimate."*

ACTIVE FILTERS AND OP-AMP CIRCUITS (E7G)

An integrated circuit operational amplifier (op-amp) is a high gain, direct-coupled differential amplifier with very high input impedance and very low output impedance. *Hint: It operates at high gain.*

The typical output impedance of an integrated circuit op-amp is very low. *Hint: Low impedance means lots of current which means lots of power and that is what you want in an amplifier.*
The typical input impedance of an integrated circuit op-amp is very high. *Hint: High input impedance means it does not load down the circuit and only draws a little power on the input.*

The op-amp offset voltage is the differential line voltage needed to bring the open loop output

voltage to zero. *Hint: You adjust the offset to bring the device to zero.*

The effect of ringing in a filter is undesired oscillations added to the desired signal. *Hint: The question is about effect not cause. The wrong answer mentions a cause.*

To prevent ringing and audio instability in a multi-section op-amp, restrict both gain and Q. *Hint: Too narrow a filter will ring and too much gain will cause instability.*

An appropriate use of an op-amp active filter is as an audio filter in a receiver.

The gain of an ideal operational amplifier does not vary with frequency. *Hint: The question is about an ideal amplifier and such an amplifier should not vary with frequency.*

Figure E7-4

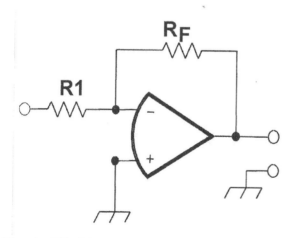

Figure E7-4 is an op-amp.

The voltage gain expected from the circuit in E7-4 when R1 is 10 ohms and RF is 470 ohms is 47. *Hint: It is the ratio of the two resistors.*

The output voltage if R1 is 1,000 ohms, RF is 10,000 ohms and 0.23 volts DC is applied to the input would be -2.3 volts. *Hint: The ratio of the resistors is 10 so the gain is 10 times but the signal is on the minus side.*

The absolute voltage gain expected when R1 is 1,800 ohms and RF is 68 kilohms is 68,000 / 1,800 = 38.

The absolute voltage gain when R1 is 3,300 ohms and RF is 47 kilohms is 47,000 / 3,300 = 14.

OSCILLATORS AND SIGNAL SOURCES (E7H)

Three oscillator circuits used in Amateur Radio are Colpitts, Hartley and Pierce. *Hint: Colpitts is only in the correct answer. Pick the Pitts. One or more of these three circuits is in all the oscillator answers.*

Positive feedback in a Hartley oscillator is supplied through a tapped coil. *Cheat: HarTley and Tapped coil.*

Positive feedback in a Colpitts oscillator is supplied through a capacitive divider. *Cheat: Colpitts and capacitive.*

Positive feedback in a Pierce oscillator is supplied through a quartz crystal. *Cheat: Mind your Ps and Qs.*

The oscillator circuits used in VFOs are Colpitts and Hartley. *Hint: Pierce is crystal, so it wouldn't do for a VFO. Find the answer with no piercings.*

A microphonic is a change in oscillator frequency due to mechanical vibration.

You can reduce an oscillator's microphonic responses by mechanically isolating the oscillator from its enclosure. *Hint: Keep it from mechanical vibration by mechanically isolating it.*

Components to reduce thermal drift in a crystal oscillator are NPO capacitors. NPO capacitors are made from ceramic and are very stable so they don't change value when heated.

The frequency synthesizer that uses a phase accumulator, lookup table, digital to analog converter and a low-pass anti-alias filter is a

direct digital synthesizer. *Hint: It has a digital to analog converter so it must be a digital synthesizer.*

The information contained in the lookup table of a direct digital synthesizer is the amplitude values that represent a sine-wave output. *Hint: A synthesizer has a sine-wave output and this is the only answer that mentions sine-wave output.*

The major spectral impurity components of direct digital synthesizers are spurious signals at discrete frequencies. *Hint: Spectral impurity is caused by spurious signals.*

To insure that a crystal oscillator provides the frequency specified by the manufacturer, provide the crystal with a specified parallel capacitance. A crystal may be designed for a particular frequency, but it can be "pulled" by stray capacitance. The crystal is calibrated for a specified parallel capacitance.

A technique for providing highly accurate and stable oscillators for microwave transmission and reception is:
Use a GPS signal reference.
Use a rubidium stabilized reference oscillator.
Use a temperature-controlled high Q dielectric resonator.
All of these choices are correct.
Hint: GPS and rubidium are easy to remember, and when you have confidence in 2 out of 3, you can be safe selecting "all of the above."

A phase-locked loop circuit is an electronic servo loop consisting of a phase detector, a low-pass filter, a voltage-controlled oscillator, and a stable reference oscillator. *Hint: If it is "locked," it must have a reference oscillator to lock with. Look for the answer with "reference oscillator" and ditch the rest of this bloated answer.*

PRACTICAL CIRCUITS

The functions that can be performed by a phase-locked loop are frequency synthesis and FM demodulation. *Hint: A loop implies an oscillator and only one answer has frequency synthesis.*

SIGNALS AND EMISSIONS (E8)

AC WAVEFORMS (E8A)

The process that shows a square wave is made up of a sine wave plus all its odd harmonics is Fourier analysis. Fourier analysis runs the signal through mathematical filters to extract its parts. A square wave contains the odd harmonics.

Fourier analysis shows a wave made up of sine waves of a given fundamental plus all its harmonics is a sawtooth wave. A sawtooth wave includes all harmonics.

The type of wave that has a rise time significantly faster than its fall time (or vice versa) is a sawtooth wave. *Hint: Imagine the jagged teeth on a saw.*

"Dither" in an analog to digital converter is a small amount of noise added to the input signal to allow a more precise representation of the true signal over time. Adding the noise prevents the converter from accumulating errors over time.

The approximate ratio of PEP-to-average power in a typical single-sideband phone signal is 2.5 to 1. *Hint: Peak is higher than average but not crazy higher.*

The factor that determines the PEP-to-average ratio of a single-sideband signal is the characteristics of the modulating signal. *Hint: Continuous and loud will put out more PEP power.*

The most accurate way to measure the RMS voltage of a complex waveform is by measuring the heating effect in a known resistor. A meter will bounce around. The resistor will heat no matter how complex the waveform.

A direct or flash conversion analog-to-digital converter would be useful for a software defined radio because very high speed allows digitizing high frequencies. *Hint: "direct or flash" implies high speed.*

An analog-to-digital converter with an 8 bit resolution can encode 256 levels. Solve 2^8 = 256.

The purpose of a low pass filter used in a digital-to-analog converter is to remove harmonics from the output caused by discrete analog levels generated. *Hint: A low pass filter removes harmonics. Who cares how they were caused?*

The types of information that can be conveyed using digital waveforms are:
Human speech.
Video signals.
Data.
All of these choices are correct.
Hint: Everything can be digitized, so all the choices are correct.

The advantage of digital signals instead of analog signals is digital signals can be regenerated multiple times without error.

The method commonly used to convert analog signals to digital signals is sequential sampling. *Hint: A converter samples the signal.*

MODULATION AND DEMODULATION (E8B)

The ratio between the frequency deviation of an RF carrier wave and the modulating frequency of its corresponding FM-phone signal is called the

modulation index. *Hint: A "ratio" is sometimes called an "index."*

The modulation index of an FM-phone signal having a maximum frequency deviation of 3000 Hz on either side of the carrier frequency when the modulating frequency is 1000 Hz is 3. Solve: The ratio of deviation (3000) and modulating frequency (1000) is 3.

The modulation index of an FM-phone signal having a maximum frequency swing of plus-or-minus 5 kHz when the maximum modulating frequency is 3 kHz. is 5000/3000 = 1.67

Deviation ratio is the ratio of the maximum carrier frequency deviation to the highest audio modulating frequency.

The deviation ratio of an FM-phone signal having a maximum frequency swing of plus or minus 7.5 kHz when the maximum modulation frequency is 3.5 kHz is 7.5/3.5 = 2.14. *Hint: Deviation ratio is the same thing as modulation index.*

The modulation index of a phase-modulated emission does not depend on the carrier frequency. *Hint: If it is phase modulated the carrier frequency doesn't change*

Orthogonal Frequency Division Multiplexing is a technique used for high-speed digital modes. *Hint: "Multiplexing" implies high speed.*

Orthogonal Frequency Division Multiplexing is a digital modulation technique using subcarriers at frequencies chosen to avoid intersymbol interference. *Hint: "multiplexing" means it uses additional subcarriers. Choose the answer with "subcarriers."*

SIGNALS AND EMISSIONS

Frequency division multiplexing is two or more information streams merged into a baseband, which then modulates the transmitter. *Hint: Multiplexing is two or more. The word to look for is "baseband."*

Digital time division multiplexing is two or more signals arranged to share discrete time slots of a data transmission. *Hint: Multiplexing "shares." Look for "time division" and "time slots."*

DIGITAL SIGNALS (E8C)

Forward Error Correction is implemented by transmitting extra data that may be used to detect and correct transmission errors. *Hint: The purpose of correction is to detect and correct errors.*

The symbol rate in a digital transmission is the rate at which the waveform of a transmitted signal changes to convey information. *Hint: Look for "rate" in the question and answer.*

The relationship of symbol rate and baud is that they are the same.

When performing phase shift keying, it is advantageous to shift phase precisely at the zero point of the crossing of the RF carriers because this results in the least possible transmitted bandwidth for the particular mode. *Hint: "Least possible bandwidth" is always good.*

The technique used to minimize bandwidth requirements of a PSK31 signal is use of sinusoidal data pulses. *Hint: Sinusoidal means a smooth and repetitive oscillation. "Smooth" sounds like it would minimize bandwidth.*

The necessary bandwidth of a 13-WPM international Morse code transmission would be 52 Hz.

The necessary bandwidth of a 170-hertz shift 300 baud ASCII transmission would be .5k Hz.

The necessary bandwidth of a 4800-Hz frequency shift 9600 baud ASCII FM transmission is 15.36 kHz.

Cheat: The answer is always about 4 times the shift or WPM:
13 x 4 = 52
170 x 4 = about 500
4800 x 4 = about 15.36 kHz.

The way ARQ accomplishes error correction is if errors are detected, retransmission is requested. *Hint: ARQ stands for Automatic Repeat reQuest.*

The digital code where each preceding or following character changes by only one bit is called the Gray code.

The advantage of Gray code in digital communications where symbols are transmitted as multiple bits is it facilitates error correction.

KEYING DEFECTS AND OVERMODULATION OF DIGITAL SIGNALS (E8D)

How the spread spectrum technique of frequency hopping works is the frequency of the transmitted signal is changed very rapidly according to a particular sequence also used by the receiving station. *Hint: They hop around with the receiver following the transmitter.*

SIGNALS AND EMISSIONS

Spread spectrum signals are resistant to interference because signals not using the spread spectrum algorithm are suppressed in the receiver. *Hint: If a signal doesn't behave like it is expected to, the receiver won't follow it.*

The spread spectrum communications technique that uses a high-speed binary bit stream to shift the phase of an RF carrier is called direct sequence. Instead of changing frequency, a binary bit stream of extra data is injected directly into the signal.

The primary effect of extremely short rise and fall times on a CW signal is the generation of key clicks.
The most common method of reducing key clicks is to increase the keying waveform rise and fall times.

A common cause of overmodulating AFSK signals is excessive transmit audio levels. *Hint: AFSK is audio frequency shift keying. If you overdrive the audio, you will overmodulate the signal.*

The parameter that might indicate that excessive high audio levels are causing distortion to an AFSK signal is the Intermodulation Distortion (IMD). *Hint: "Distortion" is in both the question and answer.*

A good minimum IMD level for an idling PSK signal is -30dB.

The difference between Baudot digital code and ASCII are Baudot uses a 5 data bits per character, and ASCII uses 7 or 8. *Hint: The answer is longer, but all you need to recognize is that Baudot uses 5 data bits.*

An advantage to using ASCII for data transmissions is it is possible to transmit both lower and upper case. Baudot is all caps with no lower case letters.

The advantage of using a parity bit with an ASCII stream is some types of errors can be detected. The parity bit tells the receiver if the ASCII stream had an even or odd number of data bits. It is the simplest form of error detection.

ANTENNAS AND TRANSMISSION LINES (E9)

BASIC ANTENNA PARAMETERS (E9A)

An isotropic antenna in a theoretical antenna used as a reference for antenna gain.
The antenna that has no gain in any direction is an isotropic antenna. *Hint: The word "isotropic" means "equal way." It is a single point that radiates equally in all directions.*

One would want to know the feed point impedance of an antenna to match impedances in order to minimize standing wave ratio on the transmission line. *Hint: You want to know impedances so you can match them.*

The factors that may affect the feed point impedance of an antenna are the antenna height, conductor length/diameter ratio and location of nearby conductive objects. *Hint: Recognize antenna height in the answer.*

The total resistance in an antenna system is the radiation resistance plus the ohmic resistance. *Hint: Ohmic resistance is resistance in the wires. Recognize "ohmic resistance" in the correct answer.*

Radiation resistance in an antenna is the value of a resistance that would dissipate the same amount of power as that radiated from an antenna.

Antenna efficiency is calculated by radiation resistance / total resistance x 100 percent. *Hint: The radiation resistance is what is radiating, the total* *resistance is everything including the ohmic resistance. The more radiating, the better.*

The factor that determines ground losses for a ground-mounted vertical antenna operating in the 3 MHz to 30 MHz range is soil conductivity. *Hint: Soil is not very conductive and therefore increases the ohmic losses.*

A way to improve the efficiency of a ground-mounted quarter-wave vertical antenna is to install a good radial system. *Hint: You want to improve on the soil conductivity.*

The beam-width of an antenna varies as the gain is increased. It becomes narrower. *Hint: The beam-width gets narrower and focuses more of the signal in one direction, increasing gain.*

Antenna gain is the ratio of the radiated signal strength of an antenna in the direction of maximum radiation to that of a reference antenna. *Hint: Gain is compared to another antenna.* Gain in an antenna is often measured in reference to a dipole dBd or isotropic dBi. A gain figure alone, without telling you what it is in reference to, is meaningless. Note it does not depend on transmitter power. The transmitter is producing the same amount of power,the antenna is concentrating it.

A half-wavelength dipole has about 2.15 dB gain over an isotropic antenna. The following questions ask you to compare the two. Just subtract 2.15 from the isotropic gain to get the dipole gain.

If an antenna has 6 dB gain over an isotropic antenna, it is 3.85 dB better than a 1/2 wavelength dipole Solve 6 – 2.15 = 3.85.

ANTENNAS AND TRANSMISSION LINES

If an antenna has 12 dB gain over an isotropic it is 9.85 dB better than a 1/2 wave dipole. Solve: 12 – 2.15 = 9.85.

The term that describes station output, taking into account all gains and losses is effective radiated power. You can determine the effective radiated power of a system by adding and subtracting the dB loss and gain of each component.

The effective radiated power relative to a dipole of a station with 150 watts transmitter power, 2 dB feed line loss, 2.2dB duplexer loss and 7 dBd of antenna gain is 286 watts. Solve: First the dBs: 7 dB -2 – 2.2 = 2.8 dBs. 3 dB of gain is double the power or 300 watts. The answer is the one that is closest to, but a little less than 300 = 286 watts.

The effective radiated power relative to a dipole of a repeater station with 200 watts transmitter output, 4 dB feed line loss, 3.2 dB duplexer loss, .08 dB circulator loss and 10 dBd of antenna gain is 317 watts. Solve: 10 – 4 – 3.2 - .8 = 2 db. 2 dB is less than double the power and the power ratio of 2 dB is not a nice even whole number so the answer is the 317.

The effective radiated power of a repeater station with 200 watts transmitter power, 2 dB feed line loss, 2.8 dB duplexer loss, 1.2 dB circulator loss and a 7 dBi antenna gain is 252 watts. Solve: The total gain is 7 – 2 – 2.8 – 1.2 = 1 dB. The answer is the one just over 200 watts.

Antenna bandwidth means the frequency range over which an antenna satisfies a performance requirement. *Hint: It tells you how wide a range the antenna can cover.* Bandwidth is often measured from one 2:1 SWR point to another centered on a frequency.

ANTENNA PATTERNS (E9B)

To determine the approximate beam-width in a given plane of a directional antenna, note the two points where the signal strength of the antenna is 3 dB less than maximum and compute the angular difference. That is what you will do in the next question.

Figure E9-1

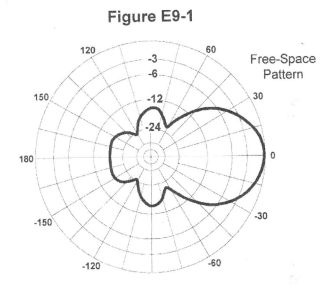

The 3 dB beam-width in the antenna radiation pattern shown in Figure E9-1 above is 50 degrees.** Solve: The pattern hits the 3 dB ring a little before 30 degress and -30 degres so the width is a little less than the total of 60 degrees. 50 degrees is the closest answer.

The front-to-back ratio is 18 dB. Solve: The front is at 0 dB, the back is at -18 dB (not well marked but it is half way between the -12 and -24 circles)

The front-to-side ratio is -14 dB. Solve: The front is 0 dB and the side is -14 dB (again, not well marked but if you look carefully you see it is closer to the -12 circle).

If a directional antenna is operated at different frequencies within the band for which it is designed, the gain may change depending on frequency. The antenna design is calculated for a particular frequency. Stray from that and the gain will change.

If a Yagi antenna is designed solely for maximum forward gain, the front-to-back decreases. *Hint: The two most important assets of a Yagi are at odds. Any design is a compromise of this conflict.*

Figure E9-2

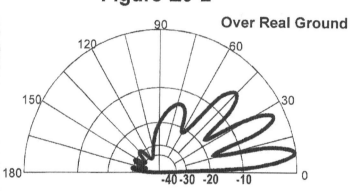

In Figure E9-2 the antenna pattern is an elevation pattern. It shows the takeoff angles.

The elevation level of the peak response is 7.5 degrees. The strongest lobe is at 7.5 degrees. This is the only question on Figure E9-2.

The total amount of radiation emitted by a directional gain antenna compared with the total amount of radiation from an isoptropic antenna assuming each is given the same amount of power, is the same. *Hint: The total amount of* radiation is the same, the directional antenna just concentrates it.

The front-to-back ratio shown in Figure E9-2 is 28 dB. It isn't well marked but you can see the back is just a little over the -30 dB line.

The number of many elevation lobes appearing in the forward direction of the antenna radiation pattern is 4. *Hint: Count the fingers.*

Antenna modeling programs use a computer program technique called Method of Moments.

The principle of a Method of Moments analysis is a wire is modeled as a series of segments, each having a uniform value of current. *Hint: Current is what causes the antenna to radiate.*

The disadvantage to decreasing the number of wire segments in an antenna model is the computed feed point impedance may be incorrect. *Hint: Fewer data points means less precision in the answer.*

The far field of an antenna is the region where the shape of the antenna pattern is independent of distance. *Hint: You are far enough away that the pattern is no longer affected by ground or objects.*

The abbreviation NEC in antenna modeling programs stands for Numerical Electromagnetic Code. *Hint: Antenna radiate electromagnetic energy. "Electromagnetic" is the key word.*

ANTENNAS AND TRANSMISSION LINES

The information obtained by submitting the details of a proposed new antenna to a modeling program can be:
SWR vs. frequency charts.
Polar plots of far-field elevation and azimuth patterns.
Antenna gain.
All of these choices are correct.
Hint: Antenna modeling programs are very powerful and provide lots of information. All of these would be capable of modeling.

WIRE AND PHASED ARRAY ANTENNAS (E9C)

The radiation pattern of two 1/4-wavelength vertical antennas spaced 1/2-wavelength apart and fed 180 degrees out of phase is a figure-8 oriented along the axis of the array.

The radiation pattern of two 1/4-wavelength vertical antennas spaced 1/4- wavelength apart and fed 90 degrees out of phase is a cardioid.

The radiation pattern of two 1/4-wavelngth vertical antennas spaced 1/2-wavelength apart and fed in phase is a figure-8 broadside to the axis of the array.

Cheat: How in the world are you supposed to remember that? Recognize:
180 oriented along.
90 cardioid.
In phase broadside.

An OCFD antenna is fed approximately 1/3 of the way from the end and with a 4:1 balun to provide multiband operation. *Hint: OCFD stands for "off-center-fed-dipole." That is all you need to recognize to answer the question.*

As the wire length is increased for an unterminated long wire antenna, the radiation pattern changes so the lobes align more in the direction of the wire. *Hint: A long wire antenna radiates in the direction of the wire.*

A rhombic antenna is two long wires pointed in the same direction in the shape of a rhombus. **The effect of a terminating resistor on a rhombic antenna is to change the radiation pattern from bidirectional to unidirectional.**

A folded dipole antenna is a dipole consisting of one-wavelength of wire forming a very thin loop.

A two-wire folded dipole antenna has a feed point impedance at the center of 300 ohms.

A G5RV antenna is multi-band dipole fed with coax and a balun through a selected length of open wire transmission line. *Hint: the G5RV is a multi-band dipole. That is all you need to know.*

A Zepp antenna is an end fed dipole. *Hint: They got their name because they were a wire trailing behind a Zeppelin and therefore, fed at the end.*

A extended double Zepp antenna is a center fed 1.25 wavelength antenna. *Hint: It is extended, so it is extra long.*

The effect on the far-field elevation pattern of a vertically polarized antenna mounted over sea water versus rocky ground is that the low-angle radiation increases. *Hint: Far off DX signals arrive at a low angle. Verticals over salt water are great low-angle DX antennas, and that is why they are used on many Dxpeditions.*

ANTENNAS AND TRANSMISSION LINES

Placing a vertical antenna over imperfect ground reduces low angle radiation. *Hint: The opposite of seawater.*

A horizontally polarized antenna mounted on a hill vs one mounted on flat ground will differ in that the takeoff angle decreases in the downhill direction. *Hint: The signal bounces off the hill further away and at a lower elevation.*

The radiation pattern of a 3-element beam antenna varies with height in that the takeoff angle decrease with increasing height. *Hint: The signal starts higher and therefore travels further before it hits the gorund. It bounces off the ground at a lower angle.*

DIRECTIONAL AND SHORTENED ANTENNAS (E9D)

The gain of an ideal parabolic dish when the operating frequency is doubled is increased by 6 dB. Solve: Gain increases with the square of the ratio of the aperture width to wavelength. Doubling the frequency increases the gain by $2^2 = 4$ times or 6 dB.

Linearly polarized Yagi antennas can be made to produce circular polarization by arranging the 2 Yagis perpendicular to each other with the driven elements at the same point fed 90 degrees out of phase. *Hint: Just remember to stack them perpendicular to each other.*

Coils are often used to lengthen a short antenna electrically. **The function of a loading coil used as part of an HF mobile antenna is to cancel positive reactance.** *Hint: Inductance cancels capacitance.*

As the frequency is lowered the feed point impedance at the base of a fixed length HF mobile antenna changes in that the radiation resistance decreases and the capacitive reactance increases. *Hint: The antenna becomes less efficient – the radiation resistance decreases. The antenna becomes too short – it has capacitive reactance.*

To minimize losses on a shortened vertical antenna, a high Q loading coil should be placed near the center of the vertical radiator.

Even better **An advantage to using top loading in a shortened vertical antenna is improved radiation efficiency** *Hint for both: The maximum current is at the base. Let it radiate before it suffers losses in the coil.*

An HF mobile antenna should have a high ratio of reactance to resistance to minimize losses. *Hint: Higher resistance would mean higher losses.*

The disadvantage of using a multiband trapped antenna is it might radiate harmonics. *Hint: It is designed to radiate on more than one band so it will.*

The bandwidth of an antenna shortened with loading coils decreases.

If the Q of an antenna increases, the SWR bandwidth decreases. *Hint: Just like in a filter, the higher Q the more selective.*

Conductors best for minimizing losses in a station's RF ground system would be wide, flat copper strap. *Hint: Skin effect means the RF flows on the surface, and the strap has more surface than round wire.*

ANTENNAS AND TRANSMISSION LINES

The best RF ground for your station would be an **electrically short connection to 3 or 4 interconnected ground rods driven into the Earth.** *Hint: Ground to the ground (Earth).*

MATCHING ANTENNAS TO FEED LINES (E9E)

The system that matches a higher impedance feed line to a lower impedance antenna by connecting the line to the driven element in two places spaced a fraction of a wavelength each side of the element center is called a **delta matching system.** *Hint: Feeding away from the center on both sides would look like a triangle (delta).*

The antenna matching system that matches an unbalanced feed line to an antenna by feeding the driven element both at the center of the element and at a fraction to one side of center is **called a gamma match.** *Hint: Fed in the center and another place is gamma.*

An effective method of connecting a 50 ohm cable to a grounded tower so it can be used as a **vertical antenna is a gamma match.**

The purpose of a series capacitor in a gamma-type matching network is to **cancel the inductive reactance of the matching network.** *Hint: Capacitors cancel inductance.*

The matching system that uses a section of transmission line connected in parallel with the feed line at or near the feed point is called **stub match.** *Hint: The parallel feedline is a stub.*

Another matching scheme is called hairpin matching. **The driven element in a 3-element Yagi tuned to use a hairpin matching system should be capacitive.** A little short.

The equivalent lumped constant network for a hairpin matching system is a shunt inductor. To make up for the capacitive element.

The term to describe the interaction at the load end of a mismatched transmission line is reflection coefficient. *Hint: A mismatched line will reflect power back.*

A measurement characteristic of a mismatched transmission line is an SWR greater than 1:1. *Hint: 1:1 is matched, so anything else is mismatched.*

An effective way to match an antenna with a 100-ohm feed point impedance to a 50-ohm coaxial cable is insert a 1/4-wavelength piece of 75-ohm coaxial cable line in series between the antenna terminals and the 50-ohm feed cable. One-quarter wavelength of coax in series can act as an impedance transformer. *Hint: Look for the answer with 1/4 wavelength in series.*

An effective way of matching a feed line to a VHF or UHF antenna when the impedances of both the antenna and feed line are unknown is to use the universal stub matching technique. *Hint: If the impedances are unknown use a universal matching technique. One size fits all.*

The purpose of a phasing line, when used in an antenna with multiple driven elements, is it insures each element operates in concert with the others to create the desired antenna pattern. *Hint: We want our antennas to act in concert.*

A use for a Wilkinson divider is to divide power equally between two 50-ohm loads while maintaining 50-ohm input impedance. *Hint: It is a divider so it divides power. The other answer about dividing frequency doesn't make any sense.*

TRANSMISSION LINES (E9F)

The term for the ratio of the actual speed at which a signal travels through a transmission line to the speed of light in a vacuum is "velocity factor."
The velocity factor of transmission line is the velocity of the wave in the transmission line divided by the velocity of light in a vacuum. *Hint: Divided by light in a vacuum. Velocity factor is never more than 1.*

The velocity factor of a transmission line is determined by the dielectric material used in the line. The type of insulation.

The physical length of a coaxial cable is shorter than its electrical length because electrical signals travel slower in transmission line than in air (or in a vacuum).

If you are cutting a line for a certain frequency, you have to take the velocity factor into account. **The velocity factor for coax is about .66.** So you would cut it to 66% of the calculated length, and the wave will arrive at the right time.

The approximate physical length of a solid polyethylene dielectric coaxial cable that is electrically one-quarter wavelength at 14.1 MHz is 3.5 meters.
Solve in your head: It is coax so the velocity factor is .66. The frequency is 20 meters so one-quarter wavelength would be 5 meters. 5 X .66 = 3.3 and 3.5 is the closest answer.
Solve with a calculator. To find the wavelength, divide 300 by the frequency. The wavelength is 300/14.1 or 21.3 meters. One-quarter would be 5.32 meters times .66 = 3.5 meters.

The approximate physical length of a solid polyethylene dielectric coaxial cable that is electrically one-quarter wavelength at 7.2 MHz is 6.9 meters.
Solve: The frequency is 40 meters so one-quarter is 10 meters. 10 x .66 = about 6.9 meters.
Solve with a calculator: The wavelength is 300/7.2 = 41.66 meters. One-quarter is 10.42. 10.42 x .66 = 6.875 meters.

The approximate physical length of an air-insulated parallel transmission line that is electrically one-half wavelength long at 14.1 MHz would be 10 meters.
Hint: Air-insulated line has no dielectric, so the velocity factor is very close to 1 (.95 actually).
Solve: The frequency is 20 meters and one-half wavelength would be 10 meters.
Solve with a calculator: We know the frequency is 21.3 meters, and half of that is 10.65. The velocity factor of the line is .95 so 10.65 x .95 = 10.1 meters

Ladder line compares to small diameter coaxial cable such as RG-58 at 50 MHz in that the ladder line has lower loss. *Hint: Ladder line has much lower losses than coax at all frequencies.*

The significant difference between foam dielectric coaxial cable and solid dielectric cable is:
Foam has lower safe operating voltage limits.
Foam has lower loss.
Foam has higher velocity factor.
All of the choices are correct.

The impedance of a 1/8 wave transmission line when the line is open at the far end is a capacitive reactance. *Hint: The center conductor and shield act like a capacitor – two separated plates.*

ANTENNAS AND TRANSMISSION LINES

The impedance of a 1/8 wave transmission line when the line is shorted at the far end is an inductive reactance. *Hint: The center conductor and shield act like an inductor. The opposite of above.*

A quarter-wave line acts as a transformer and inverts the end impedance. **The impedance of a 1/4 wavelength transmission line when the line is open at the far end is a very low impedance.**

The impedance of a 1/4 wavelength transmission line when the line is shorted at the far end is a very high impedance.

A half-wave line mirrors the end impedance. **The impedance of a 1/2 wavelength transmission line when the line is shorted at the far end is a very low impedance.** *Hint: Same as the end.*

The impedance of a 1/2 wavelength transmission line when the line is open at the far end is a very high impedance.

SMITH CHART (E9G)

Using a Smith Chart, you can calculate impedance along transmission lines. *Hint: The impedance seen at the feed point changes along the length of a transmission line as shown above. The Smith Chart is a way of computing that without using incredibly complex mathematical equations.*

A Smith Chart determines impedance and SWR values in transmission lines. *Hint: If impedance changes, so does SWR.*

The coordinate system on a Smith Chart is resistance circles and reactance arcs. *Hint: Impedance is made up of resistance and reactance.*

The two families of circles that make up a Smith Chart are resistance and reactance.

Figure E9-3

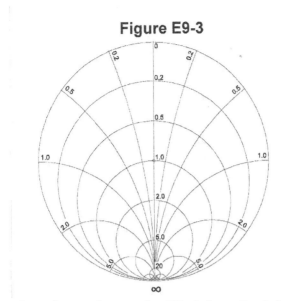

The chart shown in E9-3 is a Smith Chart.

The name of the large outer circle on which the reactance arcs terminate is the reactance axis.
Hint: It is showing the value of the reactance on that arc.

The arcs on a Smith Chart represent constant reactance. *Hint: Arcs are reactance.*

The only straight line shown is called the resistance axis. *Hint: Resistance and reactance. Reactance is arcs. Resistance is the other thing a Smith chart shows.*

The process of normalization on a Smith Chart is reassigning impedance values with regard to the prime center. *Hint: You make it normal by reassigning to the prime center.*

ANTENNAS AND TRANSMISSION LINES

The third family of circles often added to a Smith Chart during the process of solving problems are standing wave ratio circles.

The wavelength scales on a Smith Chart are calibrated in fractions of transmission line electrical wavelength. *Hint: You are calculating a transmission line depending on its length.*

RECEIVING ANTENNAS (E9H)

When constructing a Beverage antenna, the factor to include in the design to achieve good performance is it should be more than one wavelength long. *Hint: Beverages are long and low. They minimize noise on receive.*

Generally, on a low band (160 meters and 80 meters) receiving antenna, the atmospheric noise is so high that gain over a dipole is not important. *Hint: In fact, gain might make the noise worse.*

The advantage of using a shielded loop for direction finding is it is electrostatically balanced against ground, giving better nulls. *Hint: When using a direction-finding antenna, you turn it to find the null in the signal because a null is easier to hear than a peak. "Better nulls" are an advantage and all you need to recognize to answer.*

The characteristic of a cardioid pattern antenna useful for direction finding is that it has a very sharp single null. *Hint: A sharp single null is good for direction finding.*

The main drawback of a wire-loop antenna for direction finding is that it is bidirectional. *Hint: It is a drawback to hear equally well in two directions.*

The triangulation method of direction finding is antenna headings from several different receiving locations are used to locate the signal source. *Hint: Draw the headings from several receiving locations on a map and where the lines cross is the location of the signal source.*

It is advisable to use an RF attenuator on a receiver being used for direction finding because it prevents receiver overload which could make it difficult to determine peaks or nulls. *Hint: You put in enough attenuation to reduce the signal to almost nothing. Then when you point at a null, the signal will disappear completely making the null more obvious.*

A sense antenna modifies the pattern of a DF antenna array to provide a null in one direction.

A receiving loop antenna is constructed of one or more turns of wire wound in the shape of a large open coil. *Hint: A receiving loop antenna is large.*

The output voltage of a multiple turn receiving loop antenna can be increased by increasing either the number of wire turns in the loop or the area of the loop structure or both. *Hint: Bigger is better.*

SAFETY (E0A)

The primary function of an external earth connection or ground rod is lightening protection. *Hint: You want to dissipate the lightning force outside before it gets in the house.*

MPE limits are the Maximum Permissible Exposure to electrometric radiation.

When evaluating RF exposure levels from your station to your neighbor's house, you must make sure the signals from your station are less than the uncontrolled MPE limits. *Hint: Your neighbor can't control what you are doing, so he is entitled to the protection of the lower, uncontrolled limits.*

A practical way to estimate whether the RF fields produced by an amateur station are within the MPE limits would be to use an antenna modeling program to calculate field strength at accessible locations. *Hint: You want to calculate field strength.*

When evaluating a site with multiple transmitters operating at the same time, the operators of transmitters producing 5 percent or more of its total MPE limit are responsible for mitigating over-exposure situations. *Hint: Practically everyone is responsible for mitigating an unsafe situation.*

A potential hazard of using microwaves in the amateur radio bands is the high gain antennas commonly used can result in high exposure levels.

An injury from high-power UHF or microwave transmitters would be localized heating of the body from RF exposure in excess of the MPE limits.

SAR measures the rate at which RF energy is absorbed by the body. *Hint: SAR means "Soon about to roast."*

There are separate electric (E) and magnetic (H) field MPE limits because:
The body reacts to electromagnetic radiation from both fields.
Ground reflections and scattering make the field impedance vary with location.
E field and H field radiation intensity peaks can occur at different locations.
All of these choices are correct.

Dangerous levels of carbon monoxide from an emergency generator can be detected only by a carbon monoxide detector. *Hint: Carbon monoxide (CO) is colorless and odorless.*

The type of insulating material commonly used as a thermal conductor that is extremely toxic if broken or crushed and the particles are inhaled is beryllium oxide.

The toxic material that may be present in some electronic components such as high voltage capacitor and transformers is Polychlorinated Biphenyls. *Hint: You have seen them called PCBs.*

AMATEUR RADIO EXTRA CLASS QUICK SUMMARY

NOTE: The question pool is divided into Groups within Subelements. You get one question from each of the 50 Groups. If you have trouble with a Group, don't worry. You can only get one question per Group, and you may know the answer to the one you get.
The Figures (diagrams) are at the end of the text.

2016-2020 FCC Element 4 Question Pool
Effective for VEC Examinations on
July 1, 2016 thru June 30, 2020

COMMISSION'S RULES (E1)

[6 Exam Questions - 6 Groups]

E1A Operating Standards: frequency privileges; emission standards; automatic message forwarding; frequency sharing; stations aboard ships or aircraft

E1A01 [97.301, 97.305] When using a transceiver that displays the carrier frequency of phone signals, the displayed frequency that represents the highest frequency at which a properly adjusted USB emission will be totally within the band is **3 kHz below the upper band edge**

E1A02 [97.301, 97.305] When using a transceiver that displays the carrier frequency of phone signals, the displayed frequency that represents the lowest frequency at which a properly adjusted LSB emission will be totally within the band is **3 kHz above the lower band edge**

E1A03 [97.301, 97.305] With your transceiver displaying the carrier frequency of phone signals, you hear a station calling CQ on 14.349 MHz USB. Is it legal to return the call using upper sideband on the same frequency? **No, the sideband will extend beyond the band edge**

E1A04 [97.301, 97.305]
With your transceiver displaying the carrier frequency of phone signals, you hear a DX station calling CQ on 3.601 MHz LSB. Is it legal to return the call using lower sideband on the same frequency? **No, the sideband will extend beyond the edge of the phone band segment.**

E1A05 [97.313] The maximum power output permitted on the 60 meter band is **100 watts PEP effective radiated power relative to the gain of a half-wave dipole.**

E1A06 [97.15] To comply with FCC rules for 60 meter operation, the carrier frequency of a CW signal should be set **at the center frequency of the channel.**

E1A07 [97.303] The amateur band which requires transmission on specific channels rather than on a range of frequencies if the **60 meter band.**

E1A08 [97.219] If a station in a message forwarding system inadvertently forwards a message that is in violation of FCC rules, **the control operator of the originating station** is primarily accountable for the rules violation.

E1A09 [97.219] The first action you should take if your digital message forwarding station inadvertently forwards a communication that violates FCC rules is to **discontinue forwarding the communication as soon as you become aware of it.**

E1A10 [97.11] If an amateur station is installed aboard a ship or aircraft, the condition that must be met before the station is operated is **its operation must be approved by the master of the ship or the pilot in command of the aircraft.**

E1A11 [97.5] The authorization or licensing required when operating an amateur station aboard a U.S.-registered vessel in international waters is **any FCC-issued amateur license**

E1A12 [97.301, 97.305] With your transceiver displaying the carrier frequency of CW signals, you hear a DX station's CQ on 3.500 MHz, is it legal to return the call using CW on the same frequency? **No, one of the sidebands of the CW signal will be out of the band**

E1A13 [97.5] Who must be in physical control of the station apparatus of an amateur station aboard any vessel or craft that is documented or registered in the United States? **Any person holding an FCC issued amateur license or who is authorized for alien reciprocal operation**

E1A14 [97.303] The maximum bandwidth for a data emission on 60 meters is **2.8 kHz**

E1B Station restrictions and special operations: restrictions on station location; general operating restrictions, spurious emissions, control operator reimbursement; antenna structure restrictions; RACES operations; national quiet zone

E1B01 [97.3] A spurious emission is **an emission outside its necessary bandwidth that can be reduced or eliminated without affecting the information transmitted**

E1B02 [97.13] The factors which might cause the physical location of an amateur station apparatus or antenna structure to be restricted are **the location is of environmental importance or significant in American history, architecture, or culture.**

E1B03 [97.13] The distance within which an amateur station must protect an FCC monitoring facility from harmful interference is **1 mile.**

E1B04 [97.13, 1.1305-1.1319] Before placing an amateur station within an officially designated wilderness area or wildlife preserve, or an area listed in the National Register of Historical Places, **an Environmental Assessment must be submitted to the FCC.**

E1B05 [97.3] The National Radio Quiet Zone is **an area surrounding the National Radio Astronomy Observatory.**

E1B06 [97.15] The additional rules which apply if you are installing an amateur station antenna at a site at or near a public use airport are **you may have to notify the Federal Aviation Administration and register it with the FCC as required by Part 17 of FCC rules.**

E1B07 [97.307] The highest modulation index permitted at the highest modulation frequency for angle modulation below 29.0 MHz is **1.0.**

E1B08 [97.121] The limitations the FCC may place on an amateur station if its signal causes interference to domestic broadcast reception, assuming that the receivers involved are of good engineering design are **the amateur station must avoid transmitting during certain hours on frequencies that cause the interference.**

E1B09 [97.407] Amateur stations which may be operated under RACES rules are **any FCC-licensed amateur station certified by the responsible civil defense organization for the area served.**

E1B10 [97.407] Frequencies authorized to an amateur station operating under RACES rules are **all amateur service frequencies authorized to the control operator.**

SUMMARY

E1B11 [97.307] The permitted mean power of any spurious emission relative to the mean power of the fundamental emission from a station transmitter or external RF amplifier installed after January 1, 2003 and transmitting on a frequency below 30 MHZ is **at least 43 dB below.**

E1C Definitions and restrictions pertaining to local, automatic and remote control operation; control operator responsibilities for remote and automatically controlled stations; IARP and CEPT licenses; third party communications over automatically controlled stations

E1C01 [97.3] A remotely controlled station is **a station controlled indirectly through a control link**

E1C02 [97.3, 97.109] Automatic control of a station means **the use of devices and procedures for control so that the control operator does not have to be present at a control point.**

E1C03 [97.3, 97.109] The control operator responsibilities of a station under automatic control differ from one under local control in that **under automatic control the control operator is not required to be present at the control point**

E1C04 IARP is **an international amateur radio permit that allows U.S. amateurs to operate in certain countries of the Americas**

E1C05 [97.109] An automatically controlled station may originate third party communications **never.**

E1C06 [97.109] Remotely controlled amateur stations must have **a control operator must be present at the control point.**

E1C07 [97.3] Local control means **direct manipulation of the transmitter by a control operator.**

E1C08 [97.213] The maximum permissible duration of a remotely controlled station's transmissions if its control link malfunctions is **3 minutes.**

E1C09 [97.205] The range of frequencies available for an automatically controlled repeater operating below 30 MHz are **29.500 MHz - 29.700 MHz.** *Hint: Right below 30 MHz.*

E1C10 [97.113] The types of amateur stations that may automatically retransmit the radio signals of other amateur stations are **only auxiliary, repeater or space stations.**

E1C11 [97.5] The operating arrangements that allow an FCC-licensed U.S. citizen to operate in many European countries, and alien amateurs from many European countries to operate in the U.S. are the **CEPT agreement.**

E1C12 [97.117] The types of communications that may be transmitted to amateur stations in foreign countries are **communications incidental to the purpose of the amateur service and remarks of a personal nature.**

E1C13 In order to operate in accordance with CEPT rules in foreign countries where permitted, **you must bring a copy of FCC Public Notice DA 11-221.**

E1D Amateur satellites: definitions and purpose; license requirements for space stations; available frequencies and bands; telecommand and telemetry operations; restrictions, and special provisions; notification requirements

E1D01 [97.3] The term telemetry means **one-way transmission of measurements at a distance from the measuring instrument.**

SUMMARY

E1D02 [97.3] The amateur satellite service is **a radio communications service using amateur radio stations on satellites.**

E1D03 [97.3] A telecommand station in the amateur satellite service is **an amateur station that transmits communications to initiate, modify or terminate functions of a space station.**

E1D04 [97.3] An Earth station in the amateur satellite service is **an amateur station within 50 km of the Earth's surface intended for communications with amateur stations by means of objects in space.**

E1D05 [97.207] The class of licensee authorized to be the control operator of a space station is **any class with appropriate operator privileges.**

E1D06 [97.207] A requirement of a space station is **the space station must be capable of terminating transmissions by telecommand when directed by the FCC.**

E1D07 [97.207] The amateur service HF bands which have frequencies authorized for space stations are **only the 40 m, 20 m, 17 m, 15 m, 12 m and 10 m bands.**

E1D08 [97.207] The VHF amateur service bands which have frequencies available for space stations are **only 2 meters.** *Hint: VHF.*

E1D09 [97.207] The UHF amateur service bands which have frequencies available for a space station are **70 cm and 13 cm.**

E1D10 [97.211] Amateur stations eligible to be telecommand stations are **any amateur station so designated by the space station licensee, subject to the privileges of the class of operator license held by the control operator.**

E1D11 [97.209]   Amateur stations are eligible to operate as Earth stations are **any amateur station, subject to the privileges of the class of operator license held by the control operator.**

E1E Volunteer examiner program: definitions; qualifications; preparation and administration of exams; accreditation; question pools; documentation requirements

E1E01 [97.509]   The minimum number of qualified VEs required to administer an Element 4 amateur operator license examination are **3.**

E1E02 [97.523]   The questions for all written U.S. amateur license examinations listed are **in a question pool maintained by all the VECs.**

E1E03 [97.521]   A Volunteer Examiner Coordinator is **an organization that has entered into an agreement with the FCC to coordinate amateur operator license examinations.**

E1E04 [97.509, 97.525]   The Volunteer Examiner accreditation process is **the procedure by which a VEC confirms that the VE applicant meets FCC requirements to serve as an examiner.**

E1E05 [97.503]   The minimum passing score on amateur operator license examinations is **74%.**

E1E06 [97.509]   **Each administering VE** is responsible for the proper conduct and necessary supervision during an amateur operator license examination session.

E1E07 [97.509]   If a candidate fails to comply with the examiner's instructions during an amateur operator license examination, a VE should **immediately terminate the candidate's examination**

E1E08 [97.509] A VE may not administer an examination to **relatives of the VE as listed in the FCC rules.**

E1E09 [97.509] The penalty for a VE who fraudulently administers or certifies an examination is **revocation of the VE's amateur station license grant and the suspension of the VE's amateur operator license grant.**

E1E10 [97.509] After the administration of a successful examination for an amateur operator license, the administering VEs must **submit the application document to the coordinating VEC according to the coordinating VEC instructions.**

E1E11 [97.509] If an examinee scores a passing grade on all examination elements needed for an upgrade or new license, **three VEs must certify that the examinee is qualified for the license grant and that they have complied with the administering VE requirements.**

E1E12 [97.509] If the examinee does not pass the exam, the VE team must **return the application document to the examinee.**

E1E13 [97.509] If a VEC opts to conduct an exam session remotely, an acceptable form is to **use a real-time video link and the Internet to connect the exam session to the observing VEs.**

E1E14 [97.527] VEs and VECs may be reimbursed for out-of-pocket expenses **preparing, processing, administering and coordinating an examination for an amateur radio license.**

E1F Miscellaneous rules: external RF power amplifiers; business communications; compensated communications; spread spectrum; auxiliary stations; reciprocal operating privileges; special temporary authority

E1F01 [97.305] Spread spectrum transmissions are permitted **only on amateur frequencies above 222 MHz.**

E1F02 [97.107] Privileges authorized in the U.S. to persons holding an amateur service license granted by the Government of Canada are **the operating terms and conditions of the Canadian amateur service license, not to exceed U.S. Extra Class privileges.**

E1F03 [97.315] A dealer may sell an external RF power amplifier capable of operation below 144 MHz if it has not been granted FCC certification provided **it was purchased in used condition from an amateur operator and is sold to another amateur operator for use at that operator's station.**

E1F04 [97.3 "Line A" is **a line roughly parallel to and south of the U.S.-Canadian border**

E1F05 [97.303] Amateur stations located in the contiguous 48 states and north of Line A may not transmit on **420 MHz - 430 MHz.**

E1F06 [1.931] The FCC might issue a Special Temporary Authority (STA) to an amateur station to **provide for experimental amateur communications.**

E1F07 [97.113] An amateur station may send a message to a business when **neither the amateur nor his or her employer has a pecuniary interest in the communications.**

E1F08 [97.113] Prohibited Amateur station communications are **communications transmitted for hire or material compensation, except as otherwise provided in the rules.**

E1F09 [97.311] Conditions applying when transmitting spread spectrum emission are:
A station transmitting SS emission must not cause harmful interference to other stations employing other authorized emissions
The transmitting station must be in an area regulated by the FCC or in a country that permits SS emissions
The transmission must not be used to obscure the meaning of any communication
All of these choices are correct.

E1F10 [97.313] The maximum permitted transmitter peak envelope power for an amateur station transmitting spread spectrum communications is **10 W.**

E1F11 [97.317] The standards that must be met by an external RF power amplifier if it is to qualify for a grant of FCC certification are **it must satisfy the FCC's spurious emission standards when operated at the lesser of 1500 watts or its full output power.**

E1F12 [97.201] The control operator of an auxiliary station can be **only Technician, General, Advanced or Amateur Extra Class operators.** *Hint: No novices.*

OPERATING PROCEDURES (E2)

[5 Exam Questions - 5 Groups]

E2A Amateur radio in space: amateur satellites; orbital mechanics; frequencies and modes; satellite hardware; satellite operations; experimental telemetry applications

E2A01   The direction of an ascending pass for an amateur satellite is **from south to north.**

E2A02   The direction of a descending pass for an amateur satellite is **from north to south.**

E2A03   The orbital period of an Earth satellite is **the time it takes for a satellite to complete one revolution around the Earth.**

E2A04   The term mode as applied to an amateur radio satellite refers to **the satellite's uplink and downlink frequency bands.**

E2A05   The letters in a satellite's mode designator specify **the uplink and downlink frequency ranges.**

E2A06   If a satellite were operating in U/V mode, it would receive signals on **435 MHz - 438 MHz.** *Hint: the uplink is UHF so it must be 435 MHz.*

E2A07   The following types of signals can be relayed through a linear transponder:
FM and CW
SSB and SSTV
PSK and Packet
All of these choices are correct
Hint: Linear transponder can relay anything without distortion.

SUMMARY

E2A08   Effective radiated power to a satellite which uses a linear transponder should be limited **to avoid reducing the downlink power to all other users.**

E2A09   The terms L band and S band with regard to satellite communications specify **the 23 centimeter and 13 centimeter bands.**

E2A10   The received signal from an amateur satellite may exhibit a rapidly repeating fading effect **because the satellite is spinning.**

E2A11   A type of antenna that can be used to minimize the effects of spin modulation and Faraday rotation is **a circularly polarized antenna.**

E2A12   A way to predict the location of a satellite at a given time is **by calculations using the Keplerian elements for the specified satellite.**

E2A13   The type of satellite which appears to stay in one position in the sky is **geostationary.** *Hint: Stationary*

E2A14   The technology used to track, in real time, balloons carrying amateur radio transmitters is **APRS.**

E2B Television practices: fast scan television standards and techniques; slow scan television standards and techniques

E2B01   A new frame transmitted in a fast-scan (NTSC) television system **30 per second.**

E2B02   The number of horizontal lines make up a fast-scan (NTSC) television frame is **525.**

E2B03   An interlaced scanning pattern is generated in a fast-scan (NTSC) television system **by scanning odd numbered lines in one field and even numbered lines in the next.**

E2B04   Blanking in a video signal is **turning off the scanning beam while it is traveling from right to left or from bottom to top.**

E2B05   An advantage of using vestigial sideband for standard fast- scan TV transmissions is **vestigial sideband reduces bandwidth while allowing for simple video detector circuitry.**

E2B06   Vestigial sideband modulation is **amplitude modulation in which one complete sideband and a portion of the other are transmitted.**

E2B07   The name of the signal component that carries color information in NTSC video is **Chroma.**

E2B08   A common method of transmitting accompanying audio with amateur fast-scan television is:
Frequency-modulated sub-carrier
A separate VHF or UHF audio link
Frequency modulation of the video carrier
All of these choices are correct

E2B09   Other than a receiver with SSB capability and a suitable computer, **no other hardware is needed** is needed to decode SSTV using Digital Radio Mondiale (DRM).

E2B10   An acceptable bandwidth for Digital Radio Mondiale (DRM) based voice or SSTV digital transmissions made on the HF amateur bands is **3 KHz.** *Hint: same as SSB.*

E2B11   The function of the Vertical Interval Signaling (VIS) code sent as part of an SSTV transmission is **to identify the SSTV mode being used.**

E2B12   Analog SSTV images typically transmitted on the HF bands by **varying tone frequencies representing the video are transmitted using single sideband.**

E2B13   The number of lines commonly used in each frame of an amateur slow-scan color television picture are **128 or 256.**

E2B14   The aspect of an amateur slow-scan television signal that encodes the brightness of the picture is **the tone frequency.**

E2B15   **Specific tone frequencies** signal SSTV receiving equipment to begin a new picture line.

E2B16   The video standard used by North American Fast Scan ATV stations is **NTSC.**

E2B17 (B)
The approximate bandwidth of a slow-scan TV signal is **3 kHz.**

E2B18   One likely to find FM ATV transmissions on **1255 MHz.**

E2B19
Special operating frequency restrictions imposed on slow scan TV transmissions are **they are restricted to phone band segments and their bandwidth can be no greater than that of a voice signal of the same modulation type.**

E2C Operating methods: contest and DX operating; remote operation techniques; Cabrillo format; QSLing; RF network connected systems

E2C01   While contest operating, **operators are permitted to make contacts even if they do not submit a log.**

E2C02   The term self-spotting in regards to HF contest operating means **the generally prohibited practice of posting one's own call sign and frequency on a spotting network.**
Hint: Self-spotting is spotting yourself

E2C03   Contesting is generally excluded from **30 meters.**

E2C04   The type of transmission most often used for a ham radio mesh network is **spread spectrum in the 2.4 GHz band.**
Hint: A mesh network is spread around

E2C05   The function of a DX QSL Manager is **to handle the receiving and sending of confirmation cards for a DX station.**

E2C06   During a VHF/UHF contest, you expect to find the highest level of activity **in the weak signal segment of the band, with most of the activity near the calling frequency.**
Hint: People are listening for weak signals and on the calling frequency.

E2C07   The Cabrillo format is **a standard for submission of electronic contest logs.**

E2C08   Contacts which may be confirmed through the U.S. QSL bureau system are **contacts between a U.S. station and a non-U.S. station.**

E2C09   Equipment is commonly used to implement a ham radio mesh network is **a standard wireless router running custom software.**

E2C10   A DX station state that they are listening on another frequency:
Because the DX station may be transmitting on a frequency that is prohibited to some responding stations
To separate the calling stations from the DX station
To improve operating efficiency by reducing interference
All of these choices are correct

Extra Class – The Easy Way   Page 133

E2C11   You generally identify your station when attempting to contact a DX station during a contest or in a pileup by **sending your full call sign once or twice.**

E2C12   When DX signals become too weak to copy across an entire HF band a few hours after sunset, **switch to a lower frequency HF band.**
Hint: No more sunlight hurts the higher frequencies

E2C13   **No additional indicator is required to** be used by U.S.-licensed operators when operating a station via remote control where the transmitter is located in the U.S.

E2D Operating methods: VHF and UHF digital modes and procedures; APRS; EME procedures, meteor scatter procedures

E2D01   The digital modes especially designed for use for meteor scatter signals is **FSK441.**

E2D02   Good techniques for making meteor scatter contacts are:
15 second timed transmission sequences with stations alternating based on location
Use of high speed CW or digital modes
Short transmission with rapidly repeated call signs and signal report
All of these choices are correct

E2D03   The digital modes is especially useful for EME communications is **JT65.**

E2D04   The purpose of digital store-and-forward functions on an Amateur Radio satellite is **to store digital messages in the satellite for later download by other stations.**
Hint: Store and forward

E2D05   The techniques is normally used by low Earth orbiting digital satellites to relay messages around the world are **store-and-forward.**

E2D06   A method of establishing EME contacts is **time synchronous transmissions alternately from each station.**

E2D07   The digital protocol used by APRS is **AX.25.**

E2D08   The type of packet frame used to transmit APRS beacon data is **Unnumbered Information.**
Cheat: Data is information

E2D09   The digital mode with the fastest data throughput under clear communication conditions is **300 baud packet.**

E2D10   An APRS station can be used to help support a public service communications activity because **an APRS station with a GPS unit can automatically transmit information to show a mobile station's position during the event.**
Hint: APRS shows position

E2D11   The data used by the APRS network to communicate your location are **latitude and longitude.**
Hint: Latitude and longitude show location.

E2D12   JT65 improves EME communications because **it can decode signals many dB below the noise floor using FEC.** *Hint: Forward Error Correction.*

E2D13   The type of modulation used for JT65 contacts is **multi-tone AFSK.** *Hint: Audio Frequency Shift Keying*

E2D14   One advantage of using JT65 coding is **the ability to decode signals which have a very low signal to noise ratio.**

E2E Operating methods: operating HF digital modes

E2E01   The type of modulation common for data emissions below 30 MHz is **FSK.**

E2E02   The letters FEC mean as they relate to digital operation **Forward Error Correction.**

E2E03   The timing of JT65 contacts is organized by **alternating transmissions at 1 minute intervals.**

E2E04   When one of the ellipses in an FSK crossed-ellipse display suddenly disappears **selective fading has occurred.**

E2E05   A digital mode does not support keyboard-to-keyboard operation is **Winlink.**   *Hint: WINLINK is Email, not chat.*

E2E06   The most common data rate used for HF packet is **300 baud.**   *Hint: Slow keeps the signal narrow.*

E2E07   The typical bandwidth of a properly modulated MFSK16 signal is **316 Hz.**   *Hint: the 16 as in 16 Hz*

E2E08   That HF digital mode that can be used to transfer binary files is **PACTOR.**

E2E09   The HF digital mode that uses variable-length coding for bandwidth efficiency is **PSK31.**

E2E10   The digital mode with the narrowest bandwidth is **PSK31.**

E2E11   The difference between direct FSK and audio FSK is **direct FSK applies the data signal to the transmitter VFO.**

E2E12   The type of control is used by stations using the Automatic Link Enable (ALE) protocol is **automatic.** *Hint: Automatic is automatic*

E2E13   The following are possible reasons that attempts to initiate contact with a digital station on a clear frequency are unsuccessful:
Your transmit frequency is incorrect
The protocol version you are using is not the supported by the digital station
Another station you are unable to hear is using the frequency
All of these choices are correct.

RADIO WAVE PROPAGATION (E3)

[3 Exam Questions - 3 Groups]

E3A Electromagnetic waves; Earth-Moon-Earth communications; meteor scatter; microwave tropospheric and scatter propagation; aurora propagation

E3A01   The approximate maximum separation measured along the surface of the Earth between two stations communicating by Moon bounce is **12,000 miles, if the Moon is visible by both.** *Hint: The moon must be visible by both stations*

E3A02   Libration fading of an EME signal is **a fluttery irregular fading.**

E3A03   When scheduling EME contacts, the conditions that will generally result in the least path loss are **when the Moon is at perigee.** *Hint: Closest to Earth.*

E3A04   Hepburn maps predict **probability of tropospheric propagation.**

E3A05   Tropospheric propagation of microwave signals often occurs along the weather related structure of **warm and cold fronts.** *Hint: Remember tropospheric ducting from the Technician test?*

E3A06   Microwave propagation via rain scatter requires **the rain must be within radio range of both stations.**

E3A07   Atmospheric ducts capable of propagating microwave signals often form over **bodies of water.**

E3A08   When a meteor strikes the Earth's atmosphere, a cylindrical region of free electrons is formed in the **E layer.**

E3A09   The frequency range most suited for meteor scatter communications is **28 MHz - 148 MHz.**

E3A10   The type of atmospheric structure that can create a path for microwave propagation is **temperature inversion.**

E3A11   Atypical range for tropospheric propagation of microwave signals is **100 miles to 300 miles.**

E3A12   Auroral activity is caused by **the interaction in the E layer of charged particles from the Sun with the Earth's magnetic field.**

E3A13   The emission mode best for aurora propagation is **CW.**

E3A14   From the contiguous 48 states, the approximate direction an antenna should be pointed to take maximum advantage of aurora propagation is **North.**

E3A15   An electromagnetic wave is **a wave consisting of an electric field and a magnetic field oscillating at right angles to each other.**
Hint: From the Technician test- electric and magnetic

E3A16   Electromagnetic waves traveling in free space are best described as **changing electric and magnetic fields propagate the energy.**

E3A17   Circularly polarized electromagnetic waves are **waves with a rotating electric field.**

E3B Transequatorial propagation; long path; gray-line; multi-path; ordinary and extraordinary waves; chordal hop, sporadic E mechanisms

E3B01   Transequatorial propagation is **propagation between two mid-latitude points at approximately the same distance north and south of the magnetic equator.** *Hint: Across the equator.*

E3B02   The approximate maximum range for signals using transequatorial propagation is **5000 miles.**

E3B03   The best time of day for transequatorial propagation is **afternoon or early evening.**

E3B04   The terms extraordinary and ordinary waves mean **independent waves created in the ionosphere that are elliptically polarized.**

E3B05   The amateur bands which typically support long-path propagation are **160 meters to 10 meters.**
Cheat: The two extremes make a long path.

E3B06   The amateur band that most frequently provides long-path propagation is **20 meters.**

E3B07   Hearing an echo on the received signal of a distant station could be **receipt of a signal by more than one path.**
Hint: More than one signal takes more than one path

E3B08   The type of HF propagation probably occurring if radio signals travel along the terminator between daylight and darkness is **gray-line.**

E3B09   Sporadic E propagation most likely to occur **around the solstices, especially the summer solstice.**
Hint: Sporadic E is a summer-time occurrence.

E3B10   The cause of gray-line propagation is **at twilight and sunrise, D-layer absorption is low while E-layer and F-layer propagation remains high.**

E3B11   Sporadic-E propagation most likely to occur during the day **at any time.**

E3B12   The primary characteristic of chordal hop propagation is **successive ionospheric reflections without an intermediate reflection from the ground.**

E3B13   Chordal hop propagation is desirable because **the signal experiences less loss along the path compared to normal skip propagation.**

E3B14   Linearly polarized radio waves that split into ordinary and extraordinary waves in the ionosphere **become elliptically polarized.**

E3C Radio-path horizon; less common propagation modes; propagation prediction techniques and modeling; space weather parameters and amateur radio

E3C01   Ray tracing describes **modeling a radio wave's path through the ionosphere.**

E3C02   A rising A or K index indicates **increasing disruption of the geomagnetic field.**

E3C03   The signal paths most likely to experience high levels of absorption when the A index or K index is elevated is **polar paths.** *Hint: The magnetic field concentrates around the poles*

E3C04   The value of Bz (B sub Z) represents **direction and strength of the interplanetary magnetic field.**

E3C05   The orientation of Bz (B sub z) that increases the likelihood that incoming particles from the Sun will cause disturbed conditions is **southward.**

E3C06
The VHF/UHF radio horizon distance exceed the geometric horizon **by approximately 15 percent of the distance.**

E3C07
The greatest solar flare intensity is described as **Class X.** *Hint: X marks the (sun) spot*

E3C08   The space weather term G5 means **an extreme geomagnetic storm.**

E3C09   The intensity of an X3 flare compared to that of an X2 flare is **twice as great.**

E3C10   The 304A solar parameter measures **UV emissions at 304 angstroms, correlated to solar flux index.**

E3C11   VOACAP software models **HF propagation.**

E3C12   The maximum distance of ground-wave propagation **decreases** when the signal frequency is increased.

E3C13   The type of polarization best for ground-wave propagation is **vertical.**

E3C14   The radio-path horizon distance exceeds the geometric horizon because of **downward bending due to density variations in the atmosphere.**

E3C15   A sudden rise in radio background noise might indicate **a solar flare has occurred.**

AMATEUR PRACTICES (E4)

[5 Exam Questions - 5 Groups]

E4A Test equipment: analog and digital instruments; spectrum and network analyzers, antenna analyzers; oscilloscopes; RF measurements; computer aided measurements

E4A01   The parameter which determines the bandwidth of a digital or computer-based oscilloscope is the **sampling rate.**

E4A02   The parameters a spectrum analyzer would display on the vertical and horizontal axes are **RF amplitude and frequency.**

E4A03   The test instrument used to display spurious signals and/or intermodulation distortion products in an SSB transmitter is a **spectrum analyzer.** *Hint: You are analyzing the spectrum*

E4A04   The upper frequency limit for a computer soundcard-based oscilloscope program is determined by the **analog-to-digital conversion speed of the soundcard.**
Hint: It is based on the soundcard

E4A05   Advantages of a digital vs an analog oscilloscope are:
Automatic amplitude and frequency numerical readout
Storage of traces for future reference
Manipulation of time base after trace capture
All of these choices are correct

E4A06   The effect of aliasing in a digital or computer-based oscilloscope is **false signals are displayed.** *Hint: An alias is a false name.*

E4A07   An advantage of using an antenna analyzer compared to an SWR bridge to measure antenna SWR is **antenna analyzers do not need an external RF source.**

E4A08   The instrument best for measuring the SWR of a beam antenna would be **an antenna analyzer.**

E4A09   The highest frequency signal that can be digitized without aliasing when using a computer's soundcard input to digitize signals is **one-half the sample rate.**

E4A10   A **logic analyzer** displays multiple digital signal states simultaneously.

E4A11   A good practice when using an oscilloscope probe is to **keep the signal ground connection of the probe as short as possible.**

E4A12   An important precaution to follow when connecting a spectrum analyzer to a transmitter output is to **attenuate the transmitter output going to the spectrum analyzer.**

E4A13   Compensation of an oscilloscope probe is typically adjusted by **a square wave is displayed and the probe is adjusted until the horizontal portions of the displayed wave are as nearly flat as possible.**

E4A14   The purpose of the prescaler function on a frequency counter is **it divides a higher frequency signal so a low-frequency counter can display the input frequency.**

E4A15   An advantage of a period-measuring frequency counter over a direct-count type is **it provides improved resolution of low-frequency signals within a comparable time period.**

SUMMARY

E4B Measurement technique and limitations: instrument accuracy and performance limitations; probes; techniques to minimize errors; measurement of "Q"; instrument calibration; S parameters; vector network analyzers

E4B01   The factor that most affects the accuracy of a frequency counter **time base accuracy.**

E4B02   The advantage of using a bridge circuit to measure impedance is **it is very precise in obtaining a signal null.**

E4B03   A frequency counter with a specified accuracy of +/- 1.0 ppm reads 146,520,000 Hz. The most the actual frequency being measured could differ from the reading is **146.52 Hz.**
Hint: One part per million

E4B04   A frequency counter with a specified accuracy of +/- 0.1 ppm reads 146,520,000 Hz. The most the actual frequency being measured could differ from the reading is **14.652 Hz.**
Hint: 1/10 part per million

E4B05   A frequency counter with a specified accuracy of +/- 10 ppm reads 146,520,000 Hz. The most the actual frequency being measured could differ from the reading is **1465.20 Hz.**
Hint: 10 parts per million

E4B06   The power being absorbed by the load when a directional power meter connected between a transmitter and a terminating load reads 100 watts forward power and 25 watts reflected power is **75 watts.**

E4B07   The subscripts of S parameters represent **the port or ports at which measurements are made.**

E4B08   A characteristic of a good DC voltmeter is **high impedance input.**

E4B09   If the current reading on an RF ammeter placed in series with the antenna feed line of a transmitter increases as the transmitter is tuned to resonance, **there is more power going into the antenna.**   *Hint: More current = more power*

E4B10   A method to measure intermodulation distortion in an SSB transmitter is to **modulate the transmitter with two non-harmonically related audio frequencies and observe the RF output with a spectrum analyzer.**

E4B11   An antenna analyzer should be connected when measuring antenna resonance and feed point impedance by **connecting the antenna feed line directly to the analyzer's connector.**

E4B12   The significance of voltmeter sensitivity expressed in ohms per volt is **the full scale reading of the voltmeter multiplied by its ohms per volt rating will indicate the input impedance of the voltmeter.**   *Hint: E x R/E = R*

E4B13   The S parameter equivalent to forward gain is **S21.**   *Cheat; As a kid, you looked forward to being 21.*

E4B14   If a dip meter is too tightly coupled to a tuned circuit being checked, **a less accurate reading results.**

E4B15   A relative measurement of the Q for a series-tuned circuit could be **the bandwidth of the circuit's frequency response.**

E4B16   The S parameter which represents return loss or SWR is **S11.**   *Cheat: 1:1 is a good SWR.*

E4B17   Three test loads used to calibrate a standard RF vector network analyzer are **short circuit, open circuit, and 50 ohms.**

SUMMARY

E4C Receiver performance characteristics, phase noise, noise floor, image rejection, MDS, signal-to-noise-ratio; selectivity; effects of SDR receiver non-linearity

E4C01 An effect of excessive phase noise in the local oscillator section of a receiver is **it can cause strong signals on nearby frequencies to interfere with reception of weak signals.**

E4C02 The portions of a receiver that can be effective in eliminating image signal interference are **a front-end filter or pre-selector.**

E4C03 The term for the blocking of one FM phone signal by another, stronger FM phone signal is **capture effect.**

E4C04 The noise figure of a receiver is defined as **the ratio in dB of the noise generated by the receiver to the theoretical minimum noise.**

E4C05 The value of -174 dBm/Hz represents with regard to the noise floor of a receiver, **the theoretical noise at the input of a perfect receiver at room temperature.**

E4C06 A CW receiver with the AGC off has an equivalent input noise power density of -174 dBm/Hz. The level of an unmodulated carrier input to this receiver that would yield an audio output SNR of 0 dB in a 400 Hz noise bandwidth is **-148 dBm.**

E4C07 The MDS of a receiver represents **the minimum discernible signal.**

E4C08 An SDR receiver is overloaded when input signals exceed **the maximum count value of the analog-to-digital converter.**

E4C09   A good reason for selecting a high frequency for the design of the IF in a conventional HF or VHF communications receiver is **it is easier for front-end circuitry to eliminate image responses.**

E4C10   A desirable amount of selectivity for an amateur RTTY HF receiver is **300 Hz.**

E4C11   A desirable amount of selectivity for an amateur SSB phone receiver is **2.4 kHz.**

E4C12   An undesirable effect of using too wide a filter bandwidth in the IF section of a receiver is that **undesired signals may be heard.**

E4C13   A narrow-band roofing filter affects receiver performance in that **it improves dynamic range by attenuating strong signals near the receive frequency.**

E4C14   The transmit frequency that might generate an image response signal in a receiver tuned to 14.300 MHz and which uses a 455 kHz IF frequency is **15.210 MHz.**   *Hint: The local oscillator will be running at 14.755 MHz to generate a 455 kHz IF. But a signal at 15.210 would also mix with 14.755 resulting in 455 kHz out. Hint: A shortcut is to double the IF and add it to the tuned frequency.*

E4C15   The primary source of noise that is heard from an HF receiver with an antenna connected is usually **atmospheric noise.**

E4C16   Missing codes in an SDR receiver's analog-to-digital converter cause **distortion.**

E4C17   The largest effect on an SDR receiver's linearity is the **analog-to-digital converter sample width in bits.**

SUMMARY

E4D Receiver performance characteristics: blocking dynamic range; intermodulation and cross-modulation interference; 3rd order intercept; desensitization; preselector

E4D01   The blocking dynamic range of a receiver means **the difference in dB between the noise floor and the level of an incoming signal which will cause 1 dB of gain compression.**

E4D02   Two problems caused by poor dynamic range in a communications receiver are **cross-modulation of the desired signal and desensitization from strong adjacent signals.**

E4D03   Intermodulation interference between two repeaters can occur **when the repeaters are in close proximity and the signals mix in the final amplifier of one or both transmitters.**   *Hint: Intermodulation is signals mixing*

E4D04   **A properly terminated circulator at the output of the transmitter** may reduce or eliminate intermodulation interference in a repeater caused by another transmitter operating in close proximity.

E4D05   The transmitter frequencies that would cause an intermodulation-product signal in a receiver tuned to 146.70 MHz when a nearby station transmits on 146.52 MHz **are 146.34 MHz and 146.61 MHz.**

E4D06   The term for unwanted signals generated by the mixing of two or more signals is **intermodulation interference.**

E4D07   The most significant effect of an off-frequency signal when it is causing cross-modulation interference to a desired signal is **the off-frequency unwanted signal is heard in addition to the desired signal.**

E4D08   Intermodulation in an electronic circuit is caused by **nonlinear circuits or devices.**

E4D09   The purpose of the preselector in a communications receiver is **to increase rejection of unwanted signals.**

E4D10   The third-order intercept level of 40 dBm means with respect to receiver performance that **a pair of 40 dBm signals will theoretically generate a third-order intermodulation product with the same level as the input signals.**

E4D11   Third-order intermodulation products created within a receiver are of particular interest compared to other products because **the third-order product of two signals which are in the band of interest is also likely to be within the band.**

E4D12   The term for the reduction in receiver sensitivity caused by a strong signal near the received frequency is **desensitization.**

E4D13   Receiver desensitization can be caused by **strong adjacent channel signals.**

E4D14   A way to reduce the likelihood of receiver desensitization is to **decrease the RF bandwidth of the receiver.**

E4E Noise suppression: system noise; electrical appliance noise; line noise; locating noise sources; DSP noise reduction; noise blankers; grounding for signals

E4E01   The type of receiver noise that can often be reduced by use of a receiver noise blanker is **ignition noise.**

E4E02 The types of receiver noise that can often be reduced with a DSP noise filter are:
Broadband white noise
Ignition noise
Power line noise

SUMMARY

All of these choices are correct

E4E03   The signals a receiver noise blanker might be able to remove from desired signals are **signals which appear across a wide bandwidth.** *Hint: Blanker for blanket noise*

E4E04   Conducted and radiated noise caused by an automobile alternator can be suppressed **by connecting the radio's power leads directly to the battery and by installing coaxial capacitors in line with the alternator leads.**
Hint: Remember from your Technician test – connect to the battery.

E4E05   Noise from an electric motor can be suppressed by **installing a brute-force AC-line filter in series with the motor leads.** *Hint: Watch for an answer with brute force*

E4E06   A major cause of atmospheric static is **thunderstorms.**

E4E07   You can determine if line noise interference is being generated within your home **by turning off the AC power line main circuit breaker and listening on a battery operated radio.**

E4E08   The type of signal picked up by electrical wiring near a radio antenna is **a common-mode signal at the frequency of the radio transmitter.**

E4E09   An undesirable effect that can occur when using an IF noise blanker is **nearby signals may appear to be excessively wide even if they meet emission standards.**

E4E10   A common characteristic of interference caused by a touch controlled electrical device is
The interfering signal sounds like AC hum on an AM receiver or a carrier modulated by 60 Hz hum on a SSB or CW receiver.

The interfering signal may drift slowly across the HF spectrum.
The interfering signal can be several kHz in width and usually repeats at regular intervals across a HF band.
All of these choices are correct.

E4E11   If you are hearing combinations of local AM broadcast signals within one or more of the MF or HF ham bands, it is most likely caused by **nearby corroded metal joints are mixing and re-radiating the broadcast signals.**

E4E12   A disadvantage of using some types of automatic DSP notch-filters when attempting to copy CW signals is **a DSP filter can remove the desired signal at the same time as it removes interfering signals.**

E4E13   The cause of a loud roaring or buzzing AC line interference that comes and goes at intervals could be:
**Arcing contacts in a thermostatically controlled device.
A defective doorbell or doorbell transformer inside a nearby residence.
A malfunctioning illuminated advertising display.
All of these choices are correct**

E4E14   A type of electrical interference that might be caused by the operation of a nearby personal computer is **the appearance of unstable modulated or unmodulated signals at specific frequencies.**

E4E15   Shielded cables can radiate or receive interference because of **common mode currents on the shield and conductors.**

E4E16   Current that flows equally on all conductors of an unshielded multi-conductor cable is **common-mode current.**

ELECTRICAL PRINCIPLES (E5)

[4 Exam Questions - 4 Groups]

E5A Resonance and Q: characteristics of resonant circuits: series and parallel resonance; definitions and effects of Q; half-power bandwidth; phase relationships in reactive circuits

E5A01   The reason voltage across reactances in series can be larger than the voltage applied to them is **resonance.**

E5A02   Resonance in an electrical circuit is **the frequency at which the capacitive reactance equals the inductive reactance.**

E5A03   The magnitude of the impedance of a series RLC circuit at resonance is **approximately equal to circuit resistance.**

E5A04   The magnitude of the impedance of a circuit with a resistor, an inductor and a capacitor all in parallel, at resonance is **approximately equal to circuit resistance.**
Cheat: "Approximately equal to circuit resistance is always right at resonance.

E5A05   The magnitude of the current at the input of a series RLC circuit as the frequency goes through resonance is **maximum.**

E5A06   The magnitude of the circulating current within the components of a parallel LC circuit at resonance is **at a maximum.**

E5A07   The magnitude of the current at the input of a parallel RLC circuit at resonance is at a **minimum.**

E5A08   The phase relationship between the current through and the voltage across a series resonant circuit at resonance is **the voltage and current are in phase.**

E5A09   The Q of an RLC parallel resonant circuit is calculated as **resistance divided by the reactance of either the inductance or capacitance.**

E5A10   The Q of an RLC series resonant circuit is calculated as **reactance of either the inductance or capacitance divided by the resistance.**

E5A11   The half-power bandwidth of a parallel resonant circuit that has a resonant frequency of 7.1 MHz and a Q of 150 is **47.3 kHz** .
Solution: The bandwidth is frequency divided by Q.

E5A12   The half-power bandwidth of a parallel resonant circuit that has a resonant frequency of 3.7 MHz and a Q of 118 is **31.4 kHz.**

E5A13   An effect of increasing Q in a resonant circuit is **internal voltages and circulating currents increase.**

E5A14   The resonant frequency of a series RLC circuit if R is 22 ohms, L is 50 microhenrys and C is 40 picofarads is **3.56 MHz.**
Resonant frequency is determined by the formula $F = 1/(2\Pi \times \sqrt{LC})$. If that math is too much for you, as it is for me, there are online calculators that keep track of micros and picos for you so I don't feel bad about giving you this cheat: *The answer is always in the 80 or 40 meter ham band.*

E5A15 **Lower losses** can increase Q for inductors and capacitors.

E5A16   The resonant frequency of a parallel RLC circuit if R is 33 ohms, L is 50 microhenrys and C is 10 picofarads is **7.12 MHz.** *Cheat: Answer in the 40 meter ham band.*

E5A17   The result of increasing the Q of an impedance-matching circuit is the **matching bandwidth is decreased.**

E5B Time constants and phase relationships: RLC time constants; definition; time constants in RL and RC circuits; phase angle between voltage and current; phase angles of series RLC; phase angle of inductance vs susceptance; admittance and susceptance

E5B01   The term for the time required for the capacitor in an RC circuit to be charged to 63.2% of the applied voltage is **one time constant.**

E5B02   The term for the time it takes for a charged capacitor in an RC circuit to discharge to 36.8% of its initial voltage is **one time constant.**

E5B03   The phase angle of a reactance when it is converted to a susceptance results in **the sign is reversed.**   *int: It is converted*

E5B04   The time constant of a circuit having two 220 microfarad capacitors and two 1 megohm resistors, all in parallel is **220 seconds.**
Solve The formula is resistance times capacitance. Two 220 microfard capacitor in parallel are 440 microfards. Two 1 megaohm resistors in parallel are .5 megaohms. 440 x .5 = 220. *Cheat: you could just remember the answer is the value of the capacitor.*

E5B05   When the magnitude of a reactance is converted to a susceptance, **the magnitude of the susceptance is the reciprocal of the magnitude of the reactance**

E5B06   Susceptance is **the inverse of reactance..**

E5B07   The phase angle between the voltage across and the current through a series RLC circuit if XC is 500 ohms, R is 1 kilohm, and XL is 250 ohms is **14.0 degrees with the voltage lagging the current.** *Hint: Answer is always 14 degrees and ICE tells you the voltage is lagging the current since the circuit is capacitive.*

E5B08 The phase angle between the voltage across and the current through a series RLC circuit if XC is 100 ohms, R is 100 ohms, and XL is 75 ohms is **14 degrees with the voltage lagging the current.** *Hint: Answer is always 14 and ICE tells you voltage is lagging the current as the circuit is capacitive*

E5B09 The relationship between the current through a capacitor and the voltage across a capacitor is **current leads voltage by 90 degrees.** *Hint: ICE says voltage leads current. The degrees don't matter in the answer.*

E5B10 The relationship between the current through an inductor and the voltage across an inductor is **voltage leads current by 90 degrees.**

E5B11 The phase angle between the voltage across and the current through a series RLC circuit if XC is 25 ohms, R is 100 ohms, and XL is 50 ohms is **14 degrees with the voltage leading the current** *Hint: Answer is always 14 degrees and ELI tells you voltage is leading current as the circuit is inductive.*

E5B12 Admittance is **the inverse of impedance.** *Hint: The inverse of admitting is impeding.*

E5B13 The letter commonly used to represent susceptance is **B.**

E5C Coordinate systems and phasors in electronics: Rectangular Coordinates; Polar Coordinates; Phasors

E5C01 A capacitive reactance in rectangular notation is represented by **–jX.**

SUMMARY

E5C02   Impedances are escribed in polar coordinates **by phase angle and amplitude**

E5C03   An inductive reactance in polar coordinates is represented by **a positive phase angle**

E5C04   A capacitive reactance in polar coordinates is represented by a **a negative phase angle**

E5C05   The name of the diagram used to show the phase relationship between impedances at a given frequency is a **Phasor diagram.**   *Hint: Phase angle shown by phasor diagram.*

E5C06   The impedance 50–j25 represents **50 ohms resistance in series with 25 ohms capacitive reactance.**   *Hint: Capacitive reactance is always negative.*

E5C07   A vector represents **a quantity with both magnitude and an angular component.**

E5C08   A coordinate system often used to display the phase angle of a circuit containing resistance, inductive and/or capacitive reactance is **polar coordinates.**

E5C09   When using rectangular coordinates to graph the impedance of a circuit, the horizontal axis represents the **resistive component.**

E5C10   When using rectangular coordinates to graph the impedance of a circuit, the vertical axis represents the **reactive component.**

E5C11   The two numbers that are used to define a point on a graph using rectangular coordinates represent **the coordinate values along the horizontal and vertical axes.**   *Hint: A graph has a horizontal and vertical axis.*

E5C12   The plot of the impedance of a circuit using the rectangular coordinate system where the impedance point falls on the right side of the graph on the horizontal axis **is equivalent to a pure resistance.** *Hint: If it falls on the horizontal axis it no reactive component*

E5C13   The coordinate system often used to display the resistive, inductive, and/or capacitive reactance components of impedance is **rectangular coordinates.**

E5C14   The point on Figure E5-2 that best represents the impedance of a series circuit consisting of a 400 ohm resistor and a 38 picofarad capacitor at 14 MHz is **Point 4.**
Solve: We know the answer lies on the 400 ohm line of the horizontal axis (resistance). We know the circuit is capacitive so it will have a minus sign and be on the lower quadrant on the vertical axis. Point 4 it is by default!

E5C15   The point in Figure E5-2 that best represents the impedance of a series circuit consisting of a 300 ohm resistor and an 18 microhenry inductor at 3.505 MHz is **Point 3.**
Solve: Reactance is 2∏ times frequency times inductance. 2x3.14x18x3.5= 395. We know the circuit is inductive to we are looking to the top following the 300 ohm horizontal line. Point 3 is the correct answer where 300 and 395 cross.

E5C16   The point on Figure E5-2 that best represents the impedance of a series circuit consisting of a 300 ohm resistor and a 19 picofarad capacitor at 21.200 MHz is **Point 1.**
Solve: We are on the 300 ohm line of the horizontal axis and he circuit is capacitive so we must be in the lower quadrant. Point1 it is by default.

E5C17   The point on Figure E5-2 that best represents the impedance of a series circuit consisting of a 300 ohm resistor, a 0.64-microhenry inductor and an 85 picofarad capacitor at 24.900 MHz is **Point 8.**
Solve: We are on the 300 ohm line and the circuit is slightly inductive. Point 8 by default.

SUMMARY

E5D AC and RF energy in real circuits: skin effect; electrostatic and electromagnetic fields; reactive power; power factor; electrical length of conductors at UHF and microwave frequencies

E5D01   The result of skin effect is **as frequency increases, RF current flows in a thinner layer of the conductor, closer to the surface.**

E5D02   It important to keep lead lengths short for components used in circuits for VHF and above **to avoid unwanted inductive reactance.**

E5D03   Microstrip is **precision printed circuit conductors above a ground plane that provide constant impedance interconnects at microwave frequencies.**

E5D04   Short connections are necessary at microwave frequencies **to reduce phase shift along the connection.**
Hint: To reduce unwanted inductance that causes phase shift.

E5D05   The parasitic characteristic increases with conductor length is **inductance.**

E5D06   The direction in the magnetic field oriented about a conductor in relation to the direction of electron flow is **in a direction determined by the left-hand rule.**

E5D07   The strength of the magnetic field around a conductor is determined by **the amount of current flowing through the conductor.**

E5D08   The type of energy stored in an electromagnetic or electrostatic field is **potential energy.**

E5D09   Reactive power in an AC circuit that has both ideal inductors and ideal capacitors **is repeatedly exchanged between the associated magnetic and electric fields, but is**

not dissipated. *Hint: Ideal components don't dissipate power.*

E5D10   The true power can be determined in an AC circuit where the voltage and current are out of phase **by multiplying the apparent power times the power factor.**

E5D11   The power factor of an R-L circuit having a 60 degree phase angle between the voltage and the current is **0.5.** *Hint: Answer is the cosine of the degrees.*

E5D12   The watts consumed in a circuit having a power factor of 0.2 if the input is 100-VAC at 4 amperes are **80 watts.** *Hint: 100V x 4 Amps = 400 watts x .2 = 80 watts*

E5D13   The power consumed in a circuit consisting of a 100 ohm resistor in series with a 100 ohm inductive reactance drawing 1 ampere is **100 Watts.**
Foul! This is a trick question. It assumes no loss in the inductor. How much power is consumed by the resistor portion of the circuit? You are supposed to ignore the power consumed in the inductor. The answer is $P=I^2R$. I^2 is 1 times the R (100) is 100 watts.

E5D14   Reactive power is **wattless, nonproductive power.**

E5D15   The power factor of an R-L circuit having a 45 degree phase angle between the voltage and the current is **0.707.** *Hint: Answer is the cosine of the degrees.*

E5D16   The power factor of an R-L circuit having a 30 degree phase angle between the voltage and the current is **0.866.** *Hint: Answer is the cosine of the degrees.*

E5D17   The watts consumed in a circuit having a power factor of 0.6 if the input is 200VAC at 5 amperes are **600 watts.** *Hint: 200V x 5 Amps = 1000 Watts x .6 = 600 watts*

E5D18   The watts consumed in a circuit having a power factor of 0.71 if the apparent power is 500VA are **355 watts.** *Hint: 500 Watts x .71 = 355 Watts*

CIRCUIT COMPONENTS (E6)

[6 Exam Questions - 6 Groups]

E6A Semiconductor materials and devices: semiconductor materials; germanium, silicon, P-type, N-type; transistor types: NPN, PNP, junction, field-effect transistors: enhancement mode; depletion mode; MOS; CMOS; N-channel; P-channel

E6A01   Gallium arsenide is used as a semiconductor material in preference to germanium or silicon **in microwave circuits**

E6A02   Semiconductor materials containing excess free electrons are **N-type.** *Hint: Excess electrons are Negative*

E6A03   A PN-junction diode does not conduct current when reverse biased because **holes in P-type material and electrons in the N-type material are separated by the applied voltage, widening the depletion region.** *Hint: It can't conduct because of the widening region.*

E6A04   The name given to an impurity atom that adds holes to a semiconductor crystal structure is **acceptor impurity.** *Hint: It makes a hole and that accepts an electron*

E6A05   The alpha of a bipolar junction transistor is **the change of collector current with respect to emitter current.**

E6A06   The beta of a bipolar junction transistor is **the change in collector current with respect to base current.**

E6A07   An indication a silicon NPN junction transistor is biased on is **base-to-emitter voltage of approximately 0.6 to 0.7 volts.**

E6A08   The term that indicates the frequency at which the grounded-base current gain of a transistor has decreased to 0.7 of the gain obtainable at 1 kHz is the **alpha cutoff frequency.**

E6A09   A depletion-mode FET is **an FET that exhibits a current flow between source and drain when no gate voltage is applied.**

E6A10   In Figure E6-2, the schematic symbol for an N-channel dual-gate MOSFET is **4.**

E6A11   In Figure E6-2, the schematic symbol for a P-channel junction FET is **1.**

E6A12   Many MOSFET devices have internally connected Zener diodes on the gates **to reduce the chance of the gate insulation being punctured by static discharges or excessive voltages.**

E6A13   The initials CMOS stand for **Complementary Metal-Oxide Semiconductor.**

E6A14   The DC input impedance at the gate of a field-effect transistor compares with the DC input impedance of a bipolar transistor in that **an FET has high input impedance; a bipolar transistor has low input impedance.**

E6A15   The semiconductor material that contains excess holes in the outer shell of electrons is a **P-type.**

E6A16   The majority charge carriers in N-type semiconductor material are **free electrons.** *Hint: Electrons are Negative*

E6A17   The names of the three terminals of a field-effect transistor are **gate, drain, source.**

E6B Diodes

E6B01   The most useful characteristic of a Zener diode is **a constant voltage drop under conditions of varying current.** *Hint: Zener diodes are voltage regulators*

E6B02   An important characteristic of a Schottky diode as compared to an ordinary silicon diode when used as a power supply rectifier is **less forward voltage drop.**

E6B03   The special type of diode capable of both amplification and oscillation is a **tunnel diode.** *Cheat: Oscillation is an echo and tunnels echo*

E6B04   The type of semiconductor device designed for use as a voltage-controlled capacitor is a **varactor diode.** *Cheat: A varacter varies with voltage*

E6B05   The characteristic of a PIN diode that makes it useful as an RF switch or attenuator is **a large region of intrinsic material.**

E6B06   A common use of a hot-carrier diode is **as a VHF/UHF mixer or detector.**

E6B07   When a junction diode fails due to excessive current, the failure mechanism is **excessive junction temperature.** *Hint: Excessive current causes it to heat up*

E6B08   A type of semiconductor diode is described by a **metal-semiconductor junction.**

E6B09   A common use for point contact diodes is **as an RF detector.** *Cheat: You need an RF detector to make a contact*

E6B10   In Figure E6-3, the schematic symbol for a light-emitting diode is **5.** *Hint: You can see the light coming out*

E6B11 To control the attenuation of RF signals by a PIN diode use **forward DC bias current.**

E6B12 A common use for PIN diodes is **as an RF switch.**

E6B13 The type of bias is required for an LED to emit light is **forward bias.**

E6C Digital ICs: Families of digital ICs; gates; Programmable Logic Devices (PLDs)

E6C01 Te function of hysteresis in a comparator is **to prevent input noise from causing unstable output signals.**

E6C02 When the level of a comparator's input signal crosses the threshold, **the comparator changes its output state.**

E6C03 Tri-state logic is **logic devices with 0, 1, and high impedance output states.** *Hint: Three outputs states.*

E6C04 The primary advantage of tri-state logic is the ability **to connect many device outputs to a common bus.** *Hint: More choices is an advantage.*

E6C05 An advantage of CMOS logic devices over TTL devices is **lower power consumption.** *Hint; From your General test*

E6C06 CMOS digital integrated circuits have high immunity to noise on the input signal or power supply because **the input switching threshold is about one-half the power supply voltage.**

E6C07 A pull-up or pull-down resistor is best described as a **resistor connected to the positive or negative supply line used to establish a voltage when an input or output is an open circuit.**

E6C08   In Figure E6-5, the schematic symbol for a NAND gate is **2.**

E6C09   A Programmable Logic Device (PLD) is **a programmable collection of logic gates and circuits in a single integrated circuit.**

E6C10   In Figure E6-5, the schematic symbol for a NOR gate is **4.**

E6C11 In Figure E6-5, the schematic symbol for the NOT operation (inverter) is **5.**

E6C12   BiCMOS logic is **an integrated circuit logic family using both bipolar and CMOS transistors.**   *Hint: Bi and CMOS*

E6C13   An advantage of BiCMOS logic is **it has the high input impedance of CMOS and the low output impedance of bipolar transistors.**

E6C14   The primary advantage of using a Programmable Gate Array (PGA) in a logic circuit is **complex logic functions can be created in a single integrated circuit.**

E6D Toroidal and Solenoidal Inductors: permeability, core material, selecting, winding; transformers; Piezoelectric devices

E6D01 The number of turns required to produce a 5-microhenry inductor using a powdered-iron toroidal core that has an inductance index (A L) value of 40 microhenrys/100 turns is **35.**
Solve: Take the desired inductance and divide it by the turns ratio. Take the square root of that. 5/40 = .125. The square root of .125 is .353. Since the index was for 100 turns we multiple .353 by 100 to get 35.3 turns.

E6D02   The equivalent circuit of a quartz crystal is **motional capacitance, motional inductance, and loss resistance in series, all in parallel with a shunt capacitor representing electrode and stray capacitance.** *Cheat: The answer with shunt capacitance – the capacitance in the case around the crystal*

E6D03   An aspect of the piezoelectric effect is **mechanical deformation of material by the application of a voltage.**

E6D04   Materials commonly used as a slug core in a variable inductor are **ferrite and brass.**

E6D05   One reason for using ferrite cores rather than powdered-iron in an inductor is **ferrite toroids generally require fewer turns to produce a given inductance value.**

E6D06   The core material property that determines the inductance of a toroidal inductor is **permeability.**

E6D07   The usable frequency range of inductors that use toroidal cores, assuming a correct selection of core material for the frequency being used is **from less than 20 Hz to approximately 300 MHz.**

E6D08   One reason for using powdered-iron cores rather than ferrite cores in an inductor is **powdered-iron cores generally maintain their characteristics at higher currents.**

E6D09   Devices commonly used as VHF and UHF parasitic suppressors at the input and output terminals of a transistor HF amplifier are **ferrite beads.**

E6D10   A primary advantage of using a toroidal core instead of a solenoidal core in an inductor is **toroidal cores confine most of the magnetic field within the core material.** *Hint: They are a donut and confine the field.*

SUMMARY

E6D11   the number of turns required to produce a 1-mH inductor using a core that has an inductance index (A L) value of 523 millihenrys/1000 turns is **43 turns.**
Solve: 1/523 = .0019 and the square root is .0437. The index was for 1000 turns so the number of turns is 43.

E6D12   The definition of saturation in a ferrite core inductor is **the ability of the inductor's core to store magnetic energy has been exceeded.**

E6D13   The primary cause of inductor self-resonance is **inter-turn capacitance.** *Hint: To resonate, an inductor needs capacitance.*

E6D14   The type of slug material decreases inductance when inserted into a coil is **brass.** *Hint: Brass is not magnetic.*

E6D15   The current in the primary winding of a transformer if no load is attached to the secondary is called **magnetizing current.**

E6D16   The common name for a capacitor connected across a transformer secondary that is used to absorb transient voltage spikes is a **snubber capacitor.** *Cheat: It snubs the spikes*

E6D17   Core saturation of a conventional impedance matching transformer should be avoided because **harmonics and distortion could result.**

E6E Analog ICs: MMICs, CCDs, Device packages
E6E01   A charge-coupled device (CCD) **samples an analog signal and passes it in stages from the input to the output.**

E6E02   A through-hole type device is a **DIP.** *Hint: Dual Inline Package. A socket for an integrated circuit.*

E6E03   The materials likely to provide the highest frequency of operation when used in MMICs is **Gallium nitride.**

E6E04   The most common input and output impedance of circuits that use MMICs is **50 ohms.**

E6E05   The noise figure value typical of a low-noise UHF preamplifier is **2 dB.**   *Hint: It add just a little noise..*

E6E06   The characteristics of the MMIC that make it a popular choice for VHF through microwave circuits are **controlled gain, low noise figure, and constant input and output impedance over the specified frequency range.**

E6E07   To construct a MMIC-based microwave amplifier, use **microstrip construction.**

E6E08   Voltage from a power supply is normally furnished to the most common type of monolithic microwave integrated circuit (MMIC) **through a resistor and/or RF choke connected to the amplifier output lead.**

E6E09   The component package types that would be most suitable for use at frequencies above the HF range are **surface mount.**   *Hint: Very short leads.*

E6E10   The packaging technique in which leadless components are soldered directly to circuit boards is **surface mount.**

E6E11   A characteristic of DIP packaging used for integrated circuits is **a total of two rows of connecting pins placed on opposite sides of the package (Dual In-line Package).**

E6E12   High-power RF amplifier ICs and transistors are sometimes mounted in ceramic packages for **better dissipation of heat.**

E6F Optical components: photoconductive principles and effects, photovoltaic systems, optical couplers, optical sensors, and optoisolators; LCDs

E6F01   Photoconductivity is **the increased conductivity of an illuminated semiconductor.**

E6F02   When light shines on a photoconductive material, conductivity **increases.**

E6F03   The most common configuration of an optoisolator or optocoupler is **an LED and a phototransistor.** *Hint: The LED pulses on the phototransistor to pass the signal.*

E6F04   Photovoltaic effect is **the conversion of light to electrical energy.**

E6F05   An optical shaft encoder is **a device which detects rotation of a control by interrupting a light source with a patterned wheel.**

E6F06   The material affected the most by photoconductivity is **a crystalline semiconductor**

E6F07   A solid state relay is **a device that uses semiconductors to implement the functions of an electromechanical relay.**

E6F08   Optoisolators are often used in conjunction with solid state circuits when switching 120VAC because **optoisolators provide a very high degree of electrical isolation between a control circuit and the circuit being switched.**

E6F09   The efficiency of a photovoltaic cell is **the relative fraction of light that is converted to current.**

E6F10   The most common type of photovoltaic cell used for electrical power generation is **silicon.**

E6F11   The approximate open-circuit voltage produced by a fully-illuminated silicon photovoltaic cell is **0.5 V.**

E6F12   The energy from light falling on a photovoltaic cell is absorbed by **electrons.**

E6F13   A liquid crystal display (LCD) is **a display utilizing a crystalline liquid and polarizing filters which becomes opaque when voltage is applied.**

E6F14   LCD displays **may be hard view through polarized lenses.**   *Hint: LCDs depend on polarizing filters.*

PRACTICAL CIRCUITS (E7)

[8 Exam Questions - 8 Groups]

E7A Digital circuits: digital circuit principles and logic circuits: classes of logic elements; positive and negative logic; frequency dividers; truth tables

E7A01   A bi-stable circuit is **a flip-flop.**   *Hint: Bi-stable = flip or flop*

E7A02   The function of a decade counter digital IC is **it produces one output pulse for every ten input pulses.**
Hint: It counts the tens

E7A03   To divide the frequency of a pulse train by 2, use a **flip-flop**

E7A04   The number of flip-flops to divide a signal by 4 is **2.**

E7A05   The circuit that continuously alternates between two states without an external clock is an **astable multivibrator.**
Hint: It alternates and is never in one place (stable) and has no clock.

E7A06   A characteristic of a monostable multivibrator is that **it switches momentarily to the opposite binary state and then returns to its original state after a set time.**

E7A07   The logical operation an NAND gate performs is **it produces logic "0" at its output only when all inputs are logic "1."**   *Hint: It produces the opposite (N) when both (And) are the same.*

E7A08   The logical operation an OR gate performs is **it produces logic "1" at its output if any or all inputs are logic "1."**   *Hint: It answers with the same if either of the inputs are 1.*

E7A09   The logical operation performed by an exclusive NOR gate is **it produces logic "0" at its output if any single input is logic "1."**   *Hint: It gives the opposite (N) to either of the inputs (OR)*

E7A10   A truth table is **a list of inputs and corresponding outputs for a digital device.**   *Hint: A list of inputs and outputs.*

E7A11   The type of logic that defines "1" as a high voltage is **positive logic.**   *Hint: One is positive.*

E7A12   The type of logic that defines "0" as a high voltage is **negative logic.**   *Hint: 0 is negative*

E7B Amplifiers: Class of operation; vacuum tube and solid-state circuits; distortion and intermodulation; spurious and parasitic suppression; microwave amplifiers; switching-type amplifiers

E7B01   The portion of a signal cycle a Class AB amplifier operates is **more than 180 degrees but less than 360 degrees.**

E7B02   A Class D amplifier is **a type of amplifier that uses switching technology to achieve high efficiency.** *Cheat: The only answer without a D in it.*

E7B03   A component that forma the output of a class D amplifier circuit are **a low-pass filter to remove switching signal components.**

E7B04   Bias would normally be set on the load line of a Class A common emitter amplifier **approximately half-way between saturation and cutoff.** *Hint: Class A operates through the whole cycle so it has to start in the middle.*

E7B05   To prevent unwanted oscillations in an RF power amplifier, **install parasitic suppressors and/or neutralize the stage.**

E7B06   The following amplifier type that reduces or eliminates even order harmonics is **push-pull.**

E7B07   If a Class C amplifier is used to amplify a single-sideband phone s**ignal distortion and excessive bandwidth will result.** *Hint: Class C only operates a short part of the cycle so it is not linear.*

E7B08   An RF power amplifier can be neutralized **by feeding a 180-degree out-of-phase portion of the output back to the input.** *Hint: Neutralizing is to stop it from going into destructive self-oscillation.*

E7B09   The loading and tuning capacitors are to be adjusted when tuning a vacuum tube RF power amplifier that employs a Pi-network output circuit by **the tuning capacitor is adjusted for minimum plate current, and the loading capacitor is adjusted for maximum permissible plate current.** *Hint: Dip the plate and load it*

E7B10   In Figure E7-1, the purpose of R1 and R2 is **fixed bias.**

E7B11   In Figure E7-1, the purpose of R3 is **self bias.**

E7B12   The type of amplifier circuit shown in Figure E7-1 is **common emitter.**

E7B13   In Figure E7-2, the purpose of R is **emitter load.**

E7B14   Switching amplifiers are more efficient than linear amplifiers because **the power transistor is at saturation or cut off most of the time, resulting in low power dissipation.** *Hint: More efficient means lower power dissipation.*

E7B15   One way to prevent thermal runaway in a bipolar transistor amplifier is to **use a resistor in series with the emitter.**

E7B16   The effect of intermodulation products in a linear power amplifier is **transmission of spurious signals.**

E7B17   Odd-order rather than even-order intermodulation distortion products are of concern in linear power amplifiers **because they are relatively close in frequency to the desired signal.**

E7B18   A characteristic of a grounded-grid amplifier is **low input impedance.**

E7C Filters and matching networks: types of networks; types of filters; filter applications; filter characteristics; impedance matching; DSP filtering

E7C01   Capacitors and inductors of a low-pass filter Pi-network are arranged between the network's input and output as **a capacitor is connected between the input and ground, another capacitor is connected between the output and ground, and an inductor is connected between input and output.** *Hint: This would look like the Greek letter Pi*

E7C02   A property of a T-network with series capacitors and a parallel shunt inductor is **it is a high-pass filter.** *Hint: The capacitors shunt high frequencies to ground.*

E7C03   A Pi-L-network has an advantage over a regular Pi-network for impedance matching between the final amplifier of a vacuum-tube transmitter and an antenna because of **greater harmonic suppression.** *Hint: The extra L (inductor) cuts down on higher frequency harmonics.*

E7C04   An impedance-matching circuit transform a complex impedance to a resistive impedance because **it cancels the reactive part of the impedance and changes the resistive part to a desired value.**

E7C05   The filter type described as having ripple in the passband and a sharp cutoff is a **Chebyshev filter.**

E7C06   The distinguishing features of an elliptical filter are an **extremely sharp cutoff with one or more notches in the stop band.**

E7C07   A filter you would use to attenuate an interfering carrier signal while receiving an SSB transmission is a **notch filter.** *Hint: It notches out the carrier*

E7C08   The factor that has the greatest effect in helping determine the bandwidth and response shape of a crystal ladder filter is **the relative frequencies of the individual crystals.**

E7C09   A Jones filter as used as part of an HF receiver IF stage is a **variable bandwidth crystal lattice filter.**

E7C10   The filter that would be the best choice for use in a 2 meter repeater duplexer is a **cavity filter.**

E7C11   The common name for a filter network which is equivalent to two L-networks connected back-to-back with the two inductors in series and the capacitors in shunt at the input and output is a **Pi.**

E7C12   A Pi-L-network used for matching a vacuum tube final amplifier to a 50 ohm unbalanced output can be described as a **Pi-network with an additional series inductor on the output.** *Hint: Pi + L.*

E7C13   One advantage of a Pi-matching network over an L-matching network consisting of a single inductor and a single capacitor is the **Q of Pi-networks can be varied depending on the component values chosen.**

E7C14   The mode most affected by non-linear phase response in a receiver IF filter is d**igital.**

E7C15   A crystal lattice filter is a **filter with narrow bandwidth and steep skirts made using quartz crystals.** *Hint: Crystals are arranged in a lattice to get a narrower bandwidth.*

E7D Power supplies and voltage regulators; Solar array charge controllers

E7D01   One characteristic of a linear electronic voltage regulator is **the conduction of a control element is varied to maintain a constant output voltage.**

E7D02   One characteristic of a switching electronic voltage regulator **the controlled device's duty cycle is changed to produce a constant average output voltage.** *Hint: It is switched or pulsed to produce an average.*

E7D03   The device typically used as a stable reference voltage in a linear voltage regulator is a **Zener diode.**

E7D04   The linear voltage regulator that usually makes the most efficient use of the primary power source is a **series regulator.**

E7D05   The linear voltage regulator that places a constant load on the unregulated voltage source is a **shunt regulator.**

E7D06   The purpose of Q1 in the circuit shown in Figure E7-3 is **it increases the current-handling capability of the regulator.**

E7D07   The purpose of C2 in the circuit shown in Figure E7-3 is **it bypasses hum around D1.**

E7D08   The type of circuit shown in Figure E7-3 is a l**inear voltage regulator**

E7D09   The main reason to use a charge controller with a solar power system is **prevention of battery damage due to overcharge.**

E7D10   The primary reason that a high-frequency switching type high voltage power supply can be both less expensive and lighter in weight than a conventional power supply is **the high frequency inverter design uses much smaller transformers and filter components for an equivalent power output.**

E7D11   The circuit element controlled by a series analog voltage regulator to maintain a constant output voltage is a **pass transistor.**

E7D12   The drop-out voltage of an analog voltage regulator is the **minimum input-to-output voltage required to maintain regulation.**

E7D13   The equation for calculating power dissipation by a series connected linear voltage regulator is **voltage difference from input to output multiplied by output current.** *Hint: Ohm's Law: Power = Volts x Amps*

E7D14   One purpose of a "bleeder" resistor in a conventional unregulated power supply is **to improve output voltage regulation.**

E7D15   The purpose of a "step-start" circuit in a high voltage power supply is **to allow the filter capacitors to charge gradually.**

E7D16   When several electrolytic filter capacitors are connected in series to increase the operating voltage of a power supply filter circuit, resistors be connected across each capacitor:
To equalize, as much as possible, the voltage drop across each capacitor
To provide a safety bleeder to discharge the capacitors when the supply is off
To provide a minimum load current to reduce voltage excursions at light loads
All of these choices are correct

E7E Modulation and demodulation: reactance, phase and balanced modulators; detectors; mixer stages

E7E01   To generate FM phone emissions, use a **reactance modulator on the oscillator.**

E7E02   The function of a reactance modulator is **to produce PM signals by using an electrically variable inductance or capacitance.**

E7E03   An analog phase modulator functions **by varying the tuning of an amplifier tank circuit to produce PM signals.**

E7E04   One way a single-sideband phone signal can be generated is **by using a balanced modulator followed by a filter.**

E7E05   The circuit added to an FM transmitter to boost the higher audio frequencies is a **pre-emphasis network.**

E7E06   De-emphasis is commonly used in FM communications receivers **for compatibility with transmitters using phase modulation.**

E7E07   The term baseband in radio communications means **the frequency components present in the modulating signal.**

E7E08   The principal frequencies that appear at the output of a mixer circuit are **the two input frequencies along with their sum and difference frequencies.**

E7E09   When an excessive amount of signal energy reaches a mixer circuit, **spurious mixer products are generated.**

E7E10   A diode detector functions **by rectification and filtering of RF signals.**

E7E11   The type of detector used for demodulating SSB signals is a **product detector.**

E7E12   A frequency discriminator stage in a FM receiver is a **circuit for detecting FM signals**

E7F DSP filtering and other operations; Software Defined Radio Fundamentals; DSP modulation and demodulation

E7F01   Direct digital conversion as applied to software defined radios means **incoming RF is digitized by an analog-to-digital converter without being mixed with a local oscillator signal.** *Hint: The signal goes directly to the converter.*

E7F02   The kind of digital signal processing audio filter used to remove unwanted noise from a received SSB signal is an **adaptive filter.**

E7F03   The type of digital signal processing filter used to generate an SSB signal is a **Hilbert-transform filter.**

SUMMARY

E7F04   A common method of generating an SSB signal using digital signal processing is to **combine signals with a quadrature phase relationship.**

E7F05   An analog signal must be sampled by an analog-to-digital converter so that the signal can be accurately reproduced **at twice the rate of the highest frequency component of the signal.**

E7F06   The minimum number of bits required for an analog-to-digital converter to sample a signal with a range of 1 volt at a resolution of 1 millivolt is **10 bits.**

E7F07   The function Fast Fourier Transform performs is **converting digital signals from the time domain to the frequency domain.**

E7F08   The function of decimation with regard to digital filters is **reducing the effective sample rate by removing samples.**

E7F09   An anti-aliasing digital filter is required in a digital decimator because **it removes high-frequency signal components which would otherwise be reproduced as lower frequency components.**

E7F10   The of receiver analog-to-digital conversion that determines the maximum receive bandwidth of a Direct Digital Conversion SDR is the **sample rate.**

E7F11   The minimum detectable signal level for an SDR, in the absence of atmospheric or thermal noise, is set by the **reference voltage level and sample width in bits.**

E7F12   The digital process applied to I and Q signals in order to recover the baseband modulation information is **Fast Fourier Transform.**

E7F13   The function of taps in a digital signal processing filter is to **provide incremental signal delays for filter algorithms.** *Hint: More taps allow extra time for the processor to work.*

E7F14   To allow a digital signal processing filter to create a sharper filter response, use **more taps.**

E7F15   An advantage of a Finite Impulse Response (FIR) filter vs an Infinite Impulse Response (IIR) digital filter is **FIR filters delay all frequency components of the signal by the same amount.**   *Hint: Finite = defined and the same.*

E7F16   To adjust the sampling rate of an existing digital signal by a factor of 3/4, **interpolate by a factor of three, then decimate by a factor of four.**

E7F17   The letters I and Q in I/Q Modulation represent **In-phase and Quadrature.**

E7G Active filters and op-amp circuits: active audio filters; characteristics; basic circuit design; operational amplifiers

E7G01   The typical output impedance of an integrated circuit op-amp is **very low.**

E7G02   The effect of ringing in a filter is **undesired oscillations added to the desired signal.**

E7G03   The typical input impedance of an integrated circuit op-amp is **very high.**

E7G04   The term op-amp input offset voltage means **the differential input voltage needed to bring the open loop output voltage to zero.**

E7G05   To prevent unwanted ringing and audio instability in a multi-section op-amp RC audio filter circuit, **restrict both gain and Q.**

SUMMARY

E7G06   The most appropriate use of an op-amp active filter is **as an audio filter in a receiver**.

E7G07   The magnitude of voltage gain that can be expected from the circuit in Figure E7-4 when R1 is 10 ohms and RF is 470 ohms is **47**.

E7G08   The gain of an ideal operational amplifier **does not vary with frequency.**   *Hint: It is "ideal."*

E7G09   The output voltage of the circuit shown in Figure E7-4 if R1 is 1000 ohms, RF is 10,000 ohms, and 0.23 volts DC is applied to the input is **-2.3 volts.**   *Hint: The ratio of the resistors is 10 so the gain is 10 times but the signal is on the minus side.*

E7G10   The absolute voltage gain that can be expected from the circuit in Figure E7-4 when R1 is 1800 ohms and RF is 68 kilohms is **38.**

E7G11   The absolute voltage gain that can be expected from the circuit in Figure E7-4 when R1 is 3300 ohms and RF is 47 kilohms is **14.**

E7G12   An integrated circuit operational amplifier is a **high-gain, direct-coupled differential amplifier with very high input impedance and very low output impedance.**

E7H Oscillators and signal sources: types of oscillators; synthesizers and phase-locked loops; direct digital synthesizers; stabilizing thermal drift; microphonics; high accuracy oscillators

E7H01   The three oscillator circuits used in Amateur Radio equipment are **Colpitts, Hartley and Pierce.**

E7H02   A microphonic is **changes in oscillator frequency due to mechanical vibration.**

E7H03   Positive feedback is supplied in a Hartley oscillator **through a tapped coil.**

E7H04   Positive feedback is supplied in a Colpitts oscillator **through a capacitive divider.** *Hint: C = capacitive.*

E7H05   Positive feedback is supplied in a Pierce oscillator **through a quartz crystal.**

E7H06   The oscillator circuits are commonly used in VFOs are **Colpitts and Hartley.** *Hint: Not Pierce, that is crystal controlled.*

E7H07   An oscillator's microphonic responses can be reduced by **mechanically isolating the oscillator circuitry from its enclosure.**

E7H08   The components that can be used to reduce thermal drift in crystal oscillators vare **NPO capacitors.**

E7H09   The type of frequency synthesizer circuit that uses a phase accumulator, lookup table, digital to analog converter, and a low-pass anti-alias filter is a **direct digital synthesizer.**

E7H10   The information contained in the lookup table of a direct digital frequency synthesizer is **the amplitude values that represent a sine-wave output.**

E7H11   The major spectral impurity components of direct digital synthesizers are **spurious signals at discrete frequencies**

E7H12   To insure that a crystal oscillator provides the frequency specified by the crystal manufacturer, **provide the crystal with a specified parallel capacitance.**

E7H13   Techniques for providing highly accurate and stable oscillators needed for microwave transmission and reception are:

Use a GPS signal reference
Use a rubidium stabilized reference oscillator
Use a temperature-controlled high Q dielectric resonator
All of these choices are correct.

E7H14   A phase-locked loop circuit is **an electronic servo loop consisting of a phase detector, a low-pass filter, a voltage-controlled oscillator, and a stable reference oscillator.** *Hint: It needs a phase detector.*

E7H15   Functions that can be performed by a phase-locked loop are **frequency synthesis, FM demodulation.**

SIGNALS AND EMISSIONS (E8)

[4 Exam Questions - 4 Groups]

E8A AC waveforms: sine, square, sawtooth and irregular waveforms; AC measurements; average and PEP of RF signals; Fourier analysis; Analog to digital conversion: Digital to Analog conversion

E8A01   The name of the process that shows that a square wave is made up of a sine wave plus all of its odd harmonics is **Fourier analysis**

E8A02   The type of wave has a rise time significantly faster than its fall time (or vice versa) is a **sawtooth wave.**

E8A03   The type of wave a Fourier analysis shows to be made up of sine waves of a given fundamental frequency plus all of its harmonics is a **sawtooth wave.** *Cheat: "Sawtooth wave" is always the correct answer.*

E8A04   "Dither" with respect to analog to digital converters is **a small amount of noise added to the input signal to allow more precise representation of a signal over time.**

E8A05   The most accurate way of measuring the RMS voltage of a complex waveform is **by measuring the heating effect in a known resistor.**

E8A06   The approximate ratio of PEP-to-average power in a typical single-sideband phone signal is **2.5 to 1.**

E8A07   The PEP-to-average power ratio of a single-sideband phone signal is determined by **the characteristics of the modulating signal.**

E8A08   A direct or flash conversion analog-to-digital converter would be useful for a software defined radio because **very high speed allows digitizing high frequencies.**

E8A09   An analog-to-digital converter with 8 bit resolution encode **256** levels.

E8A10   The purpose of a low pass filter used in conjunction with a digital-to-analog converter is to **remove harmonics from the output caused by the discrete analog levels generated.** *Hint: A low-pass filter removes harmonics.*

E8A11   The types of information that can be conveyed using digital waveforms are:
Human speech
Video signals
Data
All of these choices are correct

E8A12   An advantage of using digital signals instead of analog signals to convey the same information is that **digital signals can be regenerated multiple times without error.**

E8A13   The method commonly used to convert analog signals to digital signals is **sequential sampling.**

SUMMARY

E8B Modulation and demodulation: modulation methods; modulation index and deviation ratio; frequency and time division multiplexing; Orthogonal Frequency Division Multiplexing

E8B01 The term for the ratio between the frequency deviation of an RF carrier wave and the modulating frequency of its corresponding FM-phone signal is the **modulation index.**

E8B02 The modulation index of a phase-modulated emission **does not depend on the RF carrier frequency.**

E8B03 The modulation index of an FM-phone signal having a maximum frequency deviation of 3000 Hz either side of the carrier frequency when the modulating frequency is 1000 Hz is **3.** Solve: $3000/1000 = 3$.

E8B04 The modulation index of an FM-phone signal having a maximum carrier deviation of plus or minus 6 kHz when modulated with a 2 kHz modulating frequency is **3.** Solve: $6/3 = 2$

E8B05 The deviation ratio of an FM-phone signal having a maximum frequency swing of plus-or-minus 5 kHz when the maximum modulation frequency is 3 kHz is **1.67.** Solve: $5/3 = 1.67$.

E8B06 The deviation ratio of an FM-phone signal having a maximum frequency swing of plus or minus 7.5 kHz when the maximum modulation frequency is 3.5 kHz is **2.14.**

E8B07 Orthogonal Frequency Division Multiplexing is a technique used for **high speed digital modes.**

E8B08 Orthogonal Frequency Division Multiplexing is **a digital modulation technique using subcarriers at frequencies chosen to avoid intersymbol interference.**

E8B09   Deviation ratio means **the ratio of the maximum carrier frequency deviation to the highest audio modulating frequency.**

E8B10   Frequency division multiplexing is **two or more information streams are merged into a baseband, which then modulates the transmitter.**

E8B11   Digital time division multiplexing **two or more signals are arranged to share discrete time slots of a data transmission.**

E8C Digital signals: digital communication modes; information rate vs bandwidth; error correction

E8C01   Forward Error Correction is implemented **by transmitting extra data that may be used to detect and correct transmission errors.**

E8C02   The symbol rate in a digital transmission is **the rate at which the waveform of a transmitted signal changes to convey information.**

E8C03   When performing phase shift keying, it is advantageous to shift phase precisely at the zero crossing of the RF carrier because **this results in the least possible transmitted bandwidth for the particular mode.**

E8C04   The technique used to minimize the bandwidth requirements of a PSK31 signal is **use of sinusoidal data pulses.**

E8C05   The necessary bandwidth of a 13-WPM international Morse code transmission is **approximately 52 Hz.**

E8C06   The necessary bandwidth of a 170-hertz shift, 300-baud ASCII transmission is **0.5 kHz.**

E8C07 The necessary bandwidth of a 4800-Hz frequency shift, 9600-baud ASCII FM transmission is **15.36 kHz.**

E8C08 ARQ accomplishes error correction by **if errors are detected, a retransmission is requested.**

E8C09 The name of a digital code where each preceding or following character changes by only one bit is **Gray code.**

E8C10 An advantage of Gray code in digital communications where symbols are transmitted as multiple bits is **it facilitates error detection.**

E8C11 The relationship between symbol rate and baud is that **they are the same.**

E8D Keying defects and overmodulation of digital signals; digital codes; spread spectrum

E8D01 Received spread spectrum signals are resistant to interference because **signals not using the spread spectrum algorithm are suppressed in the receiver.**

E8D02 The spread spectrum communications technique that uses a high speed binary bit stream to shift the phase of an RF carrier is called **direct sequence.**

E8D03 The spread spectrum technique of frequency hopping works by **the frequency of the transmitted signal is changed very rapidly according to a particular sequence also used by the receiving station.**

E8D04 The primary effect of extremely short rise or fall time on a CW signal is **the generation of key clicks.**

E8D05 The most common method of reducing key clicks is to **increase keying waveform rise and fall times.**

E8D06   Overmodulation of an AFSK signal such as PSK or MFSK is indicated by **strong ALC action.**

E8D07   A common cause of overmodulation of AFSK signals is **excessive transmit audio levels.**

E8D08   The parameter that might indicate that excessively high input levels are causing distortion in an AFSK signal is **Intermodulation Distortion (IMD).**

E8D09   A good minimum IMD level for an idling PSK signal is **-30 dB.**

E8D10   Some of the differences between the Baudot digital code and ASCII are **Baudot uses 5 data bits per character, ASCII uses 7 or 8; Baudot uses 2 characters as letters/figures shift codes, ASCII has no letters/figures shift code.** *Hint: Baudot is 5 bit is all you need to remember.*

E8D11   One advantage of using ASCII code for data communications is **it is possible to transmit both upper and lower case text.**

E8D12   The advantage of including a parity bit with an ASCII character stream is **some types of errors can be detected,**

ANTENNAS AND TRANSMISSION LINES (E9)

[8 Exam Questions - 8 Groups]

E9A Basic Antenna parameters: radiation resistance, gain, beamwidth, efficiency, beamwidth; effective radiated power, polarization

E9A01   An isotropic antenna is a **theoretical antenna used as a reference for antenna gain.**

E9A02   The antenna that has no gain in any direction is an **isotropic antenna.**

E9A03   One would need to know the feed point impedance of an antenna **to match impedances in order to minimize standing wave ratio on the transmission line.**

E9A04   Factors that may affect the feed point impedance of an antenna are **antenna height, conductor length/diameter ratio and location of nearby conductive objects.**

E9A05   The total resistance of an antenna system is **radiation resistance plus ohmic resistance.** *Hint: Total resistance must include the resistance in the wire.*

E9A06   Beam-width of an antenna varies as the gain is increased in that it **decreases.** *Hint: The signal is being squeezed narrower to increase gain in one direction.*

E9A07   Antenna gain means **the ratio of the radiated signal strength of an antenna in the direction of maximum radiation to that of a reference antenna.**

E9A08   Antenna bandwidth is **the frequency range over which an antenna satisfies a performance requirement.**

E9A09   Antenna efficiency calculated by **(radiation resistance / total resistance) x 100 per cent.**

E9A10   To improve the efficiency of a ground-mounted quarter-wave vertical antenna, **install a good radial system.**

E9A11   The factor determining ground losses for a ground-mounted vertical antenna operating in the 3 MHz to 30 MHz range is **soil conductivity.**

E9A12   The gain an antenna has compared to a 1/2-wavelength dipole when it has 6 dB gain over an isotropic antenna is **3.85 dB.**

E9A13 The gain an antenna has compared to a 1/2-wavelength dipole when it has 12 dB gain over an isotropic antenna is **9.85 dB.**

E9A14 Tthe radiation resistance of an antenna is **the value of a resistance that would dissipate the same amount of power as that radiated from an antenna.**

E9A15 The effective radiated power relative to a dipole of a repeater station with 150 watts transmitter power output, 2 dB feed line loss, 2.2 dB duplexer loss, and 7 dBd antenna gain is **286 watts.**

E9A16 The effective radiated power relative to a dipole of a repeater station with 200 watts transmitter power output, 4 dB feed line loss, 3.2 dB duplexer loss, 0.8 dB circulator loss, and 10 dBd antenna gain is **317 watts.**

E9A17 The effective radiated power of a repeater station with 200 watts transmitter power output, 2 dB feed line loss, 2.8 dB duplexer loss, 1.2 dB circulator loss, and 7 dBi antenna gain is **252 watts.**

E9A18 The term that describes station output, taking into account all gains and losses is **effective radiated power,**

E9B Antenna patterns: E and H plane patterns; gain as a function of pattern; antenna design

E9B01 The antenna radiation pattern shown in Figure E9-1, shows the 3 dB beam-width is **50 degrees.**

E9B02 The antenna radiation pattern shown in Figure E9-1, shows the front-to-back ratio is **18 dB.**

E9B03 The antenna radiation pattern shown in Figure E9-1, shows the front-to-side ratio is **14 dB.**

SUMMARY

E9B04 When a directional antenna is operated at different frequencies within the band for which it was designed, **the gain may change depending on frequency.**

E9B05 The type of antenna pattern over real ground shown in Figure E9-2 is **elevation.**

E9B06 The elevation angle of peak response in the antenna radiation pattern shown in Figure E9-2 is **7.5 degrees.**

E9B07 The total amount of radiation emitted by a directional gain antenna compares with the total amount of radiation emitted from an isotropic antenna, assuming each is driven by the same amount of power is that **they are the same.** *Hint: The total radiation is the same. The directional direction antenna focuses it.*

E9B08 To determine the approximate beam-width in a given plane of a directional antenna, **note the two points where the signal strength of the antenna is 3 dB less than maximum and compute the angular difference.** *Hint: Beam width is measured at 3 db – one half power.*

E9B09 The type of computer program technique commonly used for modeling antennas is called the **Method of Moments.**

E9B10 The principle of a Method of Moments analysis is **a wire is modeled as a series of segments, each having a uniform value of current.**

E9B11 A disadvantage of decreasing the number of wire segments in an antenna model below the guideline of 10 segments per half-wavelength is **the computed feed point impedance may be incorrect.** *Hint: Fewer data points = less accuracy.*

E9B12 The far field of an antenna is **the region where the shape of the antenna pattern is independent of distance.**

E9B13   The abbreviation NEC when applied to antenna modeling programs stands for **Numerical Electromagnetic Code.**

E9B14   The types of information can be obtained by submitting the details of a proposed new antenna to a modeling program are :
SWR vs frequency charts.
Polar plots of the far field elevation and azimuth patterns.
Antenna gain.
All of these choices are correct.

E9B15   The front-to-back ratio of the radiation pattern shown in Figure E9-2 is **28 dB.**

E9B16   The number of elevation lobes appearing in the forward direction of the antenna radiation pattern shown in Figure E9-2 is **4.**

E9C Wire and phased array antennas: rhombic antennas; effects of ground reflections; e-off angles; Practical wire antennas: Zepps, OCFD, loops

E9C01   The radiation pattern of two 1/4-wavelength vertical antennas spaced 1/2-wavelength apart and fed 180 degrees out of phase is **a figure-8 oriented along the axis of the array.**

E9C02   The radiation pattern of two 1/4 wavelength vertical antennas spaced 1/4 wavelength apart and fed 90 degrees out of phase is **cardioid.**

E9C03   The radiation pattern of two 1/4 wavelength vertical antennas spaced a 1/2 wavelength apart and fed in phase is **a figure-8 broadside to the axis of the array**

E9C04   The radiation pattern of an unterminated long wire antenna changes as the wire length is increased in that **the lobes align more in the direction of the wire.**

E9C05 An OCFD antenna is a **dipole feed approximately 1/3 the way from one end with a 4:1 balun to provide multiband operation.** *Hint: OCFD = Off Center Fed Dipole*

E9C06 The effect of a terminating resistor on a rhombic antenna is **it changes the radiation pattern from bidirectional to unidirectional.**

E9C07 The approximate feed point impedance at the center of a two-wire folded dipole antenna is **300 ohms.**

E9C08 A folded dipole antenna is a **dipole consisting of one wavelength of wire forming a very thin loop.**

E9C09   A G5RV antenna is a **multi-band dipole antenna fed with coax and a balun through a selected length of open wire transmission line.**

E9C10 A Zepp antenna is **an end-fed dipole antenna.**

E9C11 The effect on the far-field elevation pattern of a vertically polarized antenna mounted over seawater versus rocky ground is **the low-angle radiation increases.**

E9C12 An extended double Zepp antenna is a **center fed 1.25 wavelength antenna.** *Hint: It is extended, so it is longer.*

E9C13 The main effect of placing a vertical antenna over an imperfect ground is **it reduces low-angle radiation.**

E9C14 The performance of a horizontally polarized antenna mounted on the side of a hill compares with the same antenna mounted on flat ground in that **the main lobe takeoff angle decreases in the downhill direction.**

E9C15 The radiation pattern of a horizontally polarized 3-element beam antenna varies with its height above ground in that **the main lobe takeoff angle decreases with increasing height.**

E9D Directional antennas: gain; Yagi Antennas; losses; SWR bandwidth; antenna efficiency; shortened and mobile antennas; RF Grounding

E9D01   The gain of an ideal parabolic dish antenna changes when the operating frequency is doubled as **gain increases by 6 dB.**

E9D02   Linearly polarized Yagi antennas used to produce circular polarization can be made by **arrange two Yagis perpendicular to each other with the driven elements at the same point on the boom fed 90 degrees out of phase.** *Hint: Put them perpendicular so the vertical and horizontal combine.*

E9D03   To minimize losses in a shortened vertical antenna, a high Q loading coil should be placed **near the center of the vertical radiator.**

E9D04 An HF mobile antenna loading coil should have a high ratio of reactance to resistance **to minimize losses.** *Hint: Resistance burns up power as heat.*

E9D05   A disadvantage of using a multiband trapped antenna is **it might radiate harmonics.**

E9D06   The bandwidth of an antenna as it is shortened through the use of loading coils is **decreased.**

E9D07   An advantage of using top loading in a shortened HF vertical antenna is **improved radiation efficiency.**

E9D08   As the Q of an antenna increases, **SWR bandwidth decreases.** *Hint: Higher Q, more selective.*

E9D09   The function of a loading coil used as part of an HF mobile antenna is **to cancel capacitive reactance.**

E9D10 As the frequency of operation is lowered, the feed point impedance at the base of a fixed length HF mobile antenna changes in that **the radiation resistance decreases and the capacitive reactance increases.**

E9D11   A conductor that would be best for minimizing losses in a station's RF ground system is a **wide flat copper strap.** *Hint: Remember skin effect*

E9D12   The best RF ground for your station would be **an electrically short connection to 3 or 4 interconnected ground rods driven into the Earth.**

E9D13   If a Yagi antenna is designed solely for maximum forward gain, **the front-to-back ratio decreases.** *Hint: You can design for gain or front-to-back but you can't have both*

E9E Matching: matching antennas to feed lines; phasing lines; power dividers

E9E01   The system that matches a higher impedance transmission line to a lower impedance antenna by connecting the line to the driven element in two places spaced a fraction of a wavelength each side of element center is called **the delta matching system.**

E9E02   The name of an antenna matching system that matches an unbalanced feed line to an antenna by feeding the driven element both at the center of the element and at a fraction of a wavelength to one side of center is **the gamma match.**

E9E03   The name of the matching system that uses a section of transmission line connected in parallel with the feed line at or near the feed point is **the stub match.** *Hint: The line is a stub*

E9E04   The purpose of the series capacitor in a gamma-type antenna matching network is **to cancel the inductive reactance of the matching network.** *Hint: A capacitor is the opposite of an inductor.*

E9E05   In a 3-element Yagi tuned to use a hairpin matching system, **the driven element reactance must be capacitive.**

E9E06   The equivalent lumped-constant network for a hairpin matching system of a 3-element Yagi is **a shunt inductor.**

E9E07   The term that best describes the interactions at the load end of a mismatched transmission line is **reflection coefficient.**

E9E08   A characteristic of a mismatched transmission line is **an SWR greater than 1:1.**

E9E09   An effective method of connecting a 50 ohm coaxial cable feed line to a grounded tower so it can be used as a vertical antenna is a **Gamma match.**

E9E10   An effective way to match an antenna with a 100 ohm feed point impedance to a 50 ohm coaxial cable feed line is to **insert a 1/4-wavelength piece of 75 ohm coaxial cable transmission line in series between the antenna terminals and the 50 ohm feed cable.**   *Hint: Recognize the line in series as an impedance transformer.*

E9E11   An effective way of matching a feed line to a VHF or UHF antenna when the impedances of both the antenna and feed line are unknown is to **use the universal stub matching technique.**   *Hint: If the impedances are unknown use a universal matching technique. One size fits all.*

E9E12   The primary purpose of a phasing line when used with an antenna having multiple driven elements is **it ensures that each driven element operates in concert with the others to create the desired antenna pattern.**

E9E13   A Wilkinson divider **is used to divide power equally between two 50 ohm loads while maintaining 50 ohm input impedance.**

SUMMARY

E9F Transmission lines: characteristics of open and shorted feed lines; 1/8 wavelength; 1/4 wavelength; 1/2 wavelength; feed lines: coax versus open-wire; velocity factor; electrical length; coaxial cable dielectrics; velocity factor

E9F01   The velocity factor of a transmission line is **the velocity of the wave in the transmission line divided by the velocity of light in a vacuum.**

E9F02   The velocity factor of a transmission line is determined by the **dielectric materials used in the line.**

E9F03   The physical length of a coaxial cable transmission line is shorter than its electrical length because **electrical signals move more slowly in a coaxial cable than in air.**

E9F04   The typical velocity factor for a coaxial cable with solid polyethylene dielectric is **0.66.**

E9F05   The approximate physical length of a solid polyethylene dielectric coaxial transmission line that is electrically one-quarter wavelength long at 14.1 MHz is **3.5 meters.**   *Hint: It is coax so one quarter wave-length times .66*

E9F06   The approximate physical length of an air-insulated, parallel conductor transmission line that is electrically one-half wavelength long at 14.10 MHz is **10 meters.**   *Hint: It is air-insulated so it is 100%.*

E9F07   Ladder line compares to small-diameter coaxial cable such as RG-58 at 50 MHz in that it has **lower loss.**

E9F08   The term for the ratio of the actual speed at which a signal travels through a transmission line to the speed of light in a vacuum is **velocity factor.**

E9F09   The approximate physical length of a solid polyethylene dielectric coaxial transmission line that is electrically one-quarter wavelength long at 7.2 MHz is **6.9 meters.**

E9F10   The impedance a 1/8 wavelength transmission line presents to a generator when the line is shorted at the far end is **an inductive reactance.**

E9F11   The impedance a 1/8 wavelength transmission line presents to a generator when the line is open at the far end is **a capacitive reactance.**

E9F12   The impedance a 1/4 wavelength transmission line presents to a generator when the line is open at the far end is a **very low impedance.**

E9F13   The impedance a 1/4 wavelength transmission line presents to a generator when the line is shorted at the far end is a **very high impedance.** *Hint: The 1/4 wave is a transformer that inverts the impedance.*

E9F14   The impedance a 1/2 wavelength transmission line presenst to a generator when the line is shorted at the far end is a **very low impedance.** *Hint: The 1/2 wave repeats the impedance.*

E9F15   The impedance a 1/2 wavelength transmission line presents to a generator when the line is open at the far end is a **very high impedance.**

E9F16   A significant difference between foam dielectric coaxial cable and solid dielectric cable, assuming all other parameters are the same is:
Foam dielectric has lower safe operating voltage limits.
Foam dielectric has lower loss per unit of length.
Foam dielectric has higher velocity factor.
All of these choices are correct.

SUMMARY

E9G The Smith chart

E9G01 A Smith chart calculates **impedance along transmission lines.**

E9G02 The type of coordinate system used in a Smith chart is **resistance circles and reactance arcs.** *Hint: Look at a Smith chart, Figure E9-3. It is circles and arcs.*

E9G03 A Smith chart is often used to determine **impedance and SWR values in transmission lines.**

E9G04 The two families of circles and arcs that make up a Smith chart are **resistance and reactance.**

E9G05 The type of chart is shown in Figure E9-3 is a **Smith chart.**

E9G06 On the Smith chart shown in Figure E9-3, the name for the large outer circle on which the reactance arcs terminate is the **reactance axis.**

E9G07 On the Smith chart shown in Figure E9-3, the only straight line shown is **the resistance axis.**

E9G08 The process of normalization with regard to a Smith chart is the **reassigning impedance values with regard to the prime center.** *Hint: Normalizing is calibrating.*

E9G09 The third family of circles often added to a Smith chart during the process of solving problems is **standing wave ratio circles.**

E9G10 The arcs on a Smith chart represent **points with constant reactance.**

E9G11 Wavelength scales on a Smith chart are calibrated **in fractions of transmission line electrical wavelength.**

E9H Receiving Antennas: radio direction finding antennas; Beverage Antennas; specialized receiving antennas; longwire receiving antennas

E9H01 When constructing a Beverage antenna, the factor to be included in the design to achieve good performance at the desired frequency is **it should be one or more wavelengths long.** *Hint: Beverage antennas are long.*

E9H02 For low band (160 meter and 80 meter) receiving antennas, **atmospheric noise is so high that gain over a dipole is not important.**

E9H03 DELETED February 1, 2016

E9H04 An advantage of using a shielded loop antenna for direction finding is **it is electro statically balanced against ground, giving better nulls.** *Hint: Better nulls are good for direction finding.*

E9H05 The main drawback of a wire-loop antenna for direction finding is **it has a bidirectional pattern.**

E9H06 The triangulation method of direction finding is **antenna headings from several different receiving locations are used to locate the signal source.**

E9H07 It is advisable to use an RF attenuator on a receiver being used for direction finding because **it prevents receiver overload which could make it difficult to determine peaks or nulls.**

E9H08 The function of a sense antenna is **it modifies the pattern of a DF antenna array to provide a null in one direction.**

E9H09 The construction of a receiving loop antenna is **one or more turns of wire wound in the shape of a large open coil.**

E9H10   The output voltage of a multiple turn receiving loop antenna can be increased **by increasing either the number of wire turns in the loop or the area of the loop structure or both.**

E9H11   The characteristic of a cardioid pattern antenna useful for direction finding is **a very sharp single null.**

SAFETY –(E0)

[1 exam question - 1 group]

E0A Safety: amateur radio safety practices; RF radiation hazards; hazardous materials; grounding

E0A01   The primary function of an external earth connection or ground rod is **lightning protection.**

E0A02   When evaluating RF exposure levels from your station at a neighbor's home, you must **make sure signals from your station are less than the uncontrolled MPE limits.** *Hint: Uncontrolled as in "you have no control over what the recipient is doing."*

E0A03   A practical way to estimate whether the RF fields produced by an amateur radio station are within permissible MPE limits is to **use an antenna modeling program to calculate field strength at accessible locations.**

E0A04   When evaluating a site with multiple transmitters operating at the same time, the operators and licensees of which transmitters responsible for mitigating over-exposure situations are **each transmitter that produces 5 percent or more of its MPE limit at accessible locations**

E0A05   One of the potential hazards of using microwaves in the amateur radio bands is **The high gain antennas commonly used can result in high exposure levels.**

E0A06   There are separate electric (E) and magnetic (H) field MPE limits because **the body reacts to electromagnetic radiation from both the E and H fields.**
Ground reflections and scattering make the field impedance vary with location.
E field and H field radiation intensity peaks can occur at different locations.
All of these choices are correct.

E0A07   Dangerous levels of carbon monoxide from an emergency generator can be detected **only with a carbon monoxide detector.**

E0A08   SAR measures **the rate at which RF energy is absorbed by the body.** *Hint: Soon About to Roast.*

E0A09   The insulating material commonly used as a thermal conductor for some types of electronic devices is extremely toxic if broken or crushed and the particles are accidentally inhaled is **Beryllium Oxide.**

E0A10   The toxic material that may be present in some electronic components such as high voltage capacitors and transformers is **Polychlorinated Biphenyls** *Hint: PCBs.*

E0A11   Injuries that can result from using high-power UHF or microwave transmitters are **localized heating of the body from RF exposure in excess of the MPE limits.**

~ end of question pool ~

Figure E5-2

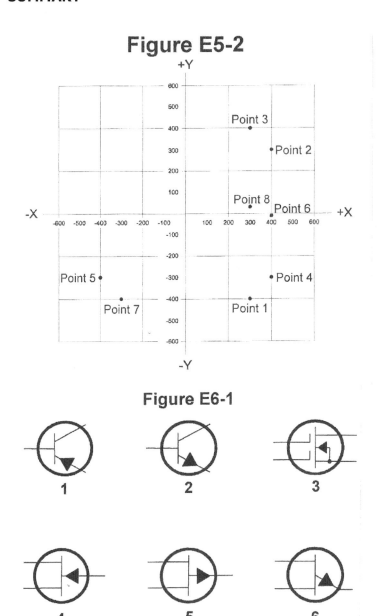

Figure E6-1

This Figure, E6-1, is NOT used on the test.

Figure E6-2

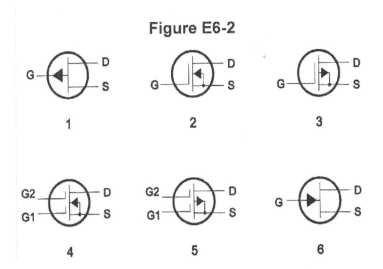

Figure E6-3

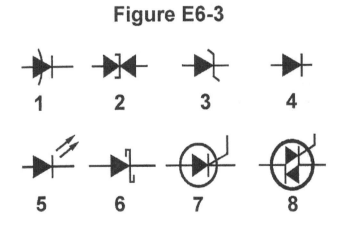

Figure E6-5

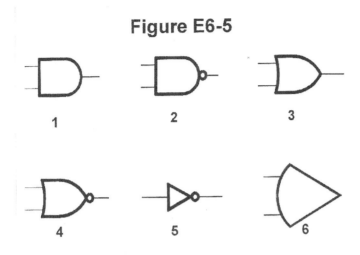

1

2

3

4

5

6

Figure E7-1

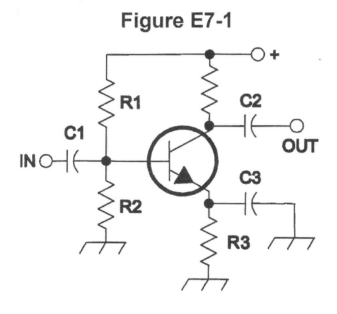

Figure E7-2

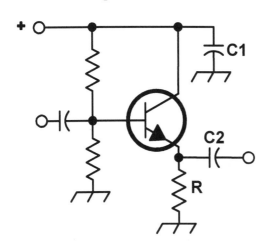

Figure E7- 3

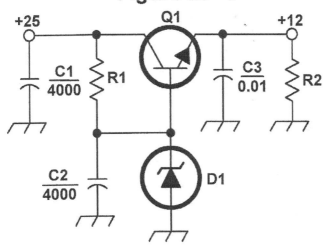

Figure E7-4

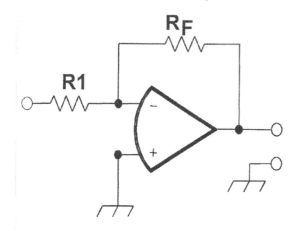

Figure E9-1

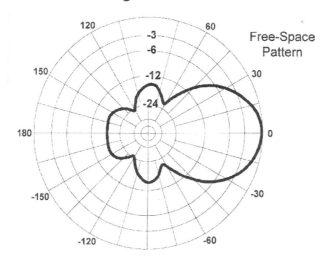

Free-Space Pattern

Figure E9-2

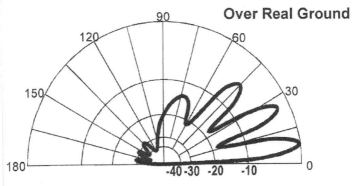

Over Real Ground

Figure E9-3

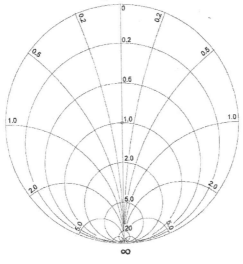

INDEX

Decimation, 84
De-emphasis, 82
delta matching, 108, 194
depletion-mode FET, 60, 161
desensitization, 44, 148, 149
Deviation ratio, 93
Digital time division multiplexing, 94
diode detector, 82, 177
DIP, 68, 166, 167
Direct digital conversion, 83
direct digital synthesizer, 89, 181
direct sequence, 96
Dither, 91
DRM, 24, 25, 131
drop-out voltage, 80, 175
DSP, 46, 47, 83, 149, 151, 172, 177
Earth station, 17, 124
effective radiated power, 12, 22, 100, 119, 130, 187, 189
elevation pattern, 102, 105, 192
elliptical filter, 77, 173
elliptically polarized, 33, 139, 140
EME, 28, 31, 134, 135, 137
far field, 103, 104, 105, 190, 191
Faraday rotation, 23, 130
Fast Fournier Transform, 84
fast-scan, 24, 25, 130, 131
ferrite cores, 66, 165

Finite Impulse Response, 85, 179
flip-flop, 72, 169
FM ATV, 25, 132
foam dielectric, 111, 197
folded dipole, 105, 192
Forward Error Correction, 94, 136, 185
Fourier analysis, 91, 182
frequency discriminator, 82, 177
front-to-back ratio, 101, 103, 189, 191, 194
front-to-side ratio, 101, 189
FSK, 29, 136
FSK441, 27, 134
G5, 35, 141
G5RV, 105, 192
Gallium arsenide, 59
gamma match, 108, 194
Gray code, 95, 186
gray-line, 33, 139, 140
grid dip meter, 41
grounded-grid, 76, 172
hairpin matching, 108, 109, 195
Hartley, 88, 180, 181
Hepburn maps, 31, 138
Hilbert-transform filter, 83, 177
hysteresis, 63, 163
IARP, 16, 122
IMD, 41, 96, 187
Infinite Impulse Response, 85, 179
interlaced scanning, 23, 131
intermodulation, 37, 41, 44, 45, 47, 76, 142,

phasor, 54, 156
Photoconductivity, 69
photovoltaic, 69, 70, 168, 169
Pierce, 88, 180, 181
piezoelectric effect, 66, 165
pi-L network, 76
PIN diode, 61, 62, 162, 163
polar coordinates, 53, 54, 156
Polychlorinated Biphenyls, 117, 201
potential energy, 56
powdered-iron, 66, 164, 165
power factor, 57, 58, 158, 159, 160
pre-emphasis, 82, 176
prescaler, 39, 143
pre-selector, 41, 146
product detector, 82
Programmable Gate Array, 65, 164
Programmable Logic Device, 64, 164
PSK31, 30, 94, 136, 185
pull-up, 64, 163
push-pull, 74
Q, 41, 49, 50, 51, 78, 84, 86, 89, 107, 144, 145, 152, 153, 154, 174, 178, 179, 182, 193
quadrature, 83, 84, 178
Quadrature, 84, 179
RACES, 15, 120, 121
Radiation resistance, 98, 188
radio horizon, 35, 141
Ray tracing, 34
reactance modulator, 81, 176
Reactive power, 57

receiving loop, 115, 199, 200
rectangular coordinates, 54, 156
rectangular notation, 53, 155
reflection coefficient, 109
resonance, 49, 50, 66, 145, 152, 153, 166
RF ground, 107, 108, 194
rhombic, 105, 191, 192
S band, 23, 130
S parameters, 40, 144
SAR, 117, 201
satellite's mode, 22
Saturation, 67
sawtooth wave, 91, 182
Schottky diode, 61, 162
SDR, 42, 46, 83, 84, 146, 147, 178
Self-spotting, 26, 133
sense antenna, 115, 199
sequential sampling, 92
series regulator, 79, 174
shielded loop, 114, 199
shunt regulator, 79, 175
silicon, 59, 70, 160, 162, 169
sinusoidal, 94, 185
skin effect, 56, 158, 194
slow-scan TV, 24, 25, 132
Smith Chart, 112
snubber capacitor, 67
soil conductivity, 99
solar flare, 35, 141, 142
solid state relay, 70, 168
Special Temporary Authority, 21, 127

spectrum analyzer, 37, 41, 142, 143, 145
Sporadic E, 34, 140
Spread spectrum, 20, 96, 133
spurious emission, 14, 120, 122, 128
SSTV, 22, 24, 25, 129, 131, 132
step-start, 81, 176
store-and-forward, 28, 134
stub match, 108, 194
surface mount, 69
Susceptance, 51
symbol rate, 94, 185, 186
takeoff angle, 106, 192
taps, 84, 85, 179
telecommand, 17, 123, 124
Telemetry, 17
thermal runaway, 75, 172
third-order intercept, 45, 149
time constant, 51, 154
T-network, 76, 173
TORROIDS, 65

transequatorial propagation, 33, 139
triangulation, 115, 199
Tri-state logic, 63
Tropospheric, 31, 138
truth table, 73, 170
tunnel diode, 61
varactor diode, 61
VEC, 18, 19, 118, 125, 126
vector, 41, 53, 144, 145, 156
vector network analyzer, 41, 145
velocity factor, 110, 111, 196, 197
Vertical Interval Signaling, 25, 131
Vestigial sideband, 24, 131
VOCAP, 35
Wilkinson divider, 109, 195
Winlink, 30, 136
wire-loop, 114, 199
Zener diode, 61, 78, 162, 174
Zepp, 105, 192

Made in the USA
Columbia, SC
05 November 2017